Building Down Ba

Building Down Barriers

A guide to construction best practice

Clive Thomas Cain

Spon Press
Taylor & Francis Group

LONDON AND NEW YORK

First published 2003
by Spon Press
11 New Fetter Lane, London EC4P 4EE

Simultaneously published in the USA and Canada
by Spon Press
29 West 35th Street, New York, NY 10001

Spon Press is an imprint of the Taylor & Francis Group

© 2003 Clive Thomas Cain

Typeset in Sabon by
Keystroke, Jacaranda Lodge, Wolverhampton
Printed and bound in Great Britain by
TJ International Ltd, Padstow, Cornwall

British Library Cataloguing in Publication Data
A catalogue record for this book is available from the British Library

Library of Congress Cataloging in Publication Data
A catalog record for this book has been requested

ISBN 0–415–28965–3

Contents

About the author

Until December 2000, Clive Thomas Cain was the Quality Director at Defence Estates with responsibility for monitoring the excellence of its outputs using the European Foundation for Quality Management 'Business Excellence Model'.

He was also responsible for the pioneering Building Down Barriers development project for which he was awarded a CBE. This developed the unique *Handbook of Supply Chain Management*, which can be used to integrate design and construction to deliver client satisfaction through a single point of responsibility – client satisfaction being the delivery of maximum functionality for the minimum cost of ownership, which includes the elimination of inefficiency and waste in the design and construction process.

The outstanding success of the Building Down Barriers project caused the handbook process to be adopted as the foundation of the three definitive best practice standards, namely the Confederation of Construction Clients' *Charter Handbook*, the National Audit Office Report, *Modernising Construction*, and the Department of Culture, Media and Sport Report, *Better Public Buildings*, commissioned by the Prime Minister, Tony Blair.

Prior to his early retirement, Clive Thomas Cain was a member of the Movement for Innovation Board and was the Defence Estates' representative at the Design Build Foundation and on the Construction Round Table. He was also actively involved with the Construction Clients' Forum (now the Confederation of Construction Clients), where he chaired the Steering Group responsible for *Whole Life Costing: A Client's Guide*.

Since December 2000, he has been coaching in supply chain management (lean construction) and has written a concise 'How To' guide entitled *Supply Chain Integration in Building Services*, published by the Chartered Institute of Building Services Engineers in the UK, and the definitive booklet, *A Guide to Best Practice in Construction Procurement*, published by the UK Construction Best Practice Programme.

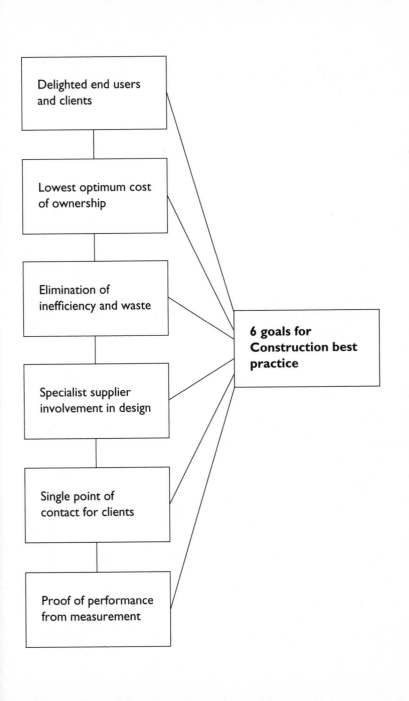

Introduction

This guide investigates how to generate higher and more assured profits for all involved in the construction procurement process. These profits will come from eliminating the unnecessary costs in the design and construction process whilst delivering a higher quality product. This book is equally valid for new construction, refurbishment and maintenance works and is aimed at everyone in the design and construction supply chain, from manufacturers to end users, and from Chief Executives to operatives. It describes what they need to do differently on Monday, why they need to do it, and what improved benefits, profits and products will flow from the change process.

It explains, in simple terms, the historical background to the current step-change in UK client procurement practice and describes the key aspects of the three best practice standards that now define the culture, the relationships and the processes of all involved in UK public and private construction procurement. The three standards are the Confederation of Construction Clients' *Charter Handbook*, published in December 2000, the National Audit Office report, *Modernising Construction*, published in January 2001, and the Department of Culture, Media and Sport report, *Better Public Buildings*, commissioned by the Prime Minister and published in October 2000.

The publication of these radical, definitive and harmonised best practice standards has, for the first time, given the UK construction industry clear and indisputable goals for the reforms which were initially demanded in the Latham Report in 1994 and confirmed by the Egan Report in 1998. Their existence means that public and private clients in the UK are now under intense external pressure to radically change the traditional, sequential procurement process that the Latham and Egan Reports and the three standards see as the primary cause of poor value in the construction industry. As a consequence, understanding the three standards and foreseeing their implications is a survival issue for UK construction industry firms, especially design consultants.

The guide builds on the experience of the joint Ministry of Defence and Department of the Environment, Transport and the Regions Building Down Barriers initiative that was launched by the Construction Minister, Nick Raynsford, in 1997 to develop a comprehensive toolset for integrating the design and construction supply chain, to test and refine it on two live pilot projects and measure the improvements. The exceptional improvements measured on the two pilot projects caused all three standards (*Charter Handbook*, *Modernising Construction*, and *Better Public Buildings*) to build on the experiences of the Building Down Barriers initiative. These experiences were thoroughly investigated by the National Audit Office and are recorded in their report *Modernising Construction*. Those wishing to understand the theory and empirical background behind this guide need to read the *Building Down Barriers Handbook of Supply Chain Management*, published by CIRIA (see Chapter 11, 'Further reading').

The guide explains, on a sector-by-sector basis, how to assess current working practices and compare them with the best practice of the three standards:

- it has structured questionnaires that can be used for first or third party assessments
- it provides advice on a best value tendering process that can be used to select an integrated design and construction team
- it provides detailed guidance on how to radically improve current practice, with the guidance differentiated between the various sectors of the industry, namely the clients (major repeat

clients as well as small and occasional clients), the consultants, the construction contractors and the specialist suppliers

- it is of as much value to small and medium enterprises as it is to major enterprises
- it is as applicable to maintenance and refurbishment as it is to new construction works
- it provides detailed guidance on the use of value management techniques within an integrated design and construction supply chain to ensure that the completed building will delight the end users and deliver the best value in whole-life performance
- it provides detailed guidance on how to select independent experts who will have the appropriate knowledge and experience to properly support the change process
- it looks at the very real commercial benefits that will come from embracing the best practice of the three UK standards and warns of the certain and fatal consequences of ignoring the six primary goals they set for the UK construction industry.

1 Seventy years of customer demand for improvement

The first major report reviewing the performance of the UK construction industry was produced in 1929 and there have been around thirteen similar reports produced between 1929 and 1994. All were inspired by client concerns about the impact on their commercial performance of the inefficiency and waste in the construction industry, and all contained remarkably similar messages. These client concerns were very effectively summed up in a book entitled *Reaching for the Skies* written by an architect called Alfred Bossom in 1934. He went to America in the early part of the last century and became closely involved in the design and construction of skyscrapers. This taught him that construction could be treated as an engineering process in which everything is scheduled in advance and all work is carried out to an agreed timetable. The result of using these engineering techniques meant that buildings were able to be erected more quickly than they were in Great Britain, yet cost no more. They yielded larger profits for both the building owner and the contractor and enabled the operatives to be paid from three to five times the wages they received in Great Britain. On his return, Bossom saw the weaknesses in the performance of the British construction industry with unblinkered eyes and became an enthusiastic advocate for radical change.

In his book he stated:

> All rents and costs of production throughout Great Britain are higher than they should be because houses and factories cost

too much and take too long to build. For the same reason the building industry languishes, employment in it is needlessly precarious and some of our greatest national needs, like the clearing away of the slums and the reconditioning of our factories, are rendered almost prohibitive on the score of expense.

The process of construction, instead of being an orderly and consecutive advance down the line, is all too apt to become a scramble and a muddle.

Bad layouts add at least 15% to the production of the cotton industry. Of how many of our steel plants and woollen mills, and even our relatively up-to-date motor works might not the same be said? The battle of trade may easily be lost before it has fairly been opened – in the architect's designing room.

This description of a fragmented, inefficient and adversarial industry in 1934, which damaged the commercial effectiveness of its end user clients by being guilty of passing on unnecessarily high capital costs and poor functionality, seems little different from that described in the Latham Report, *Constructing the Team*, in 1994 or the Egan Report, *Rethinking Construction*, in 1998. In fact, the only thing different in the 1994 and 1998 reports is the realisation that the maintenance and running costs are also unnecessarily high.

The reason why the numerous reports between 1929 and 1994 failed to have any impact on the performance of the construction industry is because the industry continues to be blind to its failings. It was also unwilling to measure its performance, particularly the impact of fragmentation and adversarial attitudes on the effective utilisation of labour and materials and the lack of effective pre-planning of construction activities that had concerned Alfred Bossom in 1934. Because clients continued to reinforce fragmentation and adversarial attitudes by insisting on using a sequential procurement process, the situation was made worse. Subsequently, it became impossible to harness the skills and knowledge of the specialist suppliers into design development because they were not involved until after the construction contractor was appointed and the design was complete. Consequently, it was impossible for them to inject buildability and 'right first time' or greater standard-isation of components into the developing design.

Fortunately, the Latham Report in 1994 proved a major catalyst in persuading clients to actively lead the reform movement, rather than standing to one side and expecting the industry to take the initiative. The reason for this radical change in client attitudes was that the Latham Report, for the first time, put a figure of 30 per cent on the cost of inefficiency and waste in the industry. Across the entire construction industry, this burden of unnecessary cost could amount to as much as £17 billion each year. Within the public sector annual expenditure of around £23 billion, it could amount to as much as £7 billion each year.

For individual repeat clients, the message about the high level of unnecessary cost was a powerful driver for them to take a much more active role in industry reform. The Latham Report led to the formation of influential client groups whose sole intent was to force the pace and direction of reform. The Construction Round Table had been formed in 1992 after the demise of NEDO (the National Economic Development Council), by a small group of major repeat clients such as BAA, McDonald's, Whitbread, Unilever and Transco. The Latham Report findings re-energised Construction Round Table members and encouraged them to take a more active and overt leadership role in the industry, which culminated in the publication of their *Agenda for Change*. The Construction Clients' Forum was formed in 1994 from a mixture of client umbrella bodies, such as the British Property Federation, and major repeat clients such as Defence Estates. The Government Construction Clients Panel was formed in 1997 to provide a single, collective voice for government procurement agencies and departments. In addition to these client groupings, pan-industry groups with dominant client leadership were also formed. The Reading Construction Forum was incor-porated in 1995, and the Design Build Foundation incorporated in 1997.

In 1998, the Egan Report strongly reinforced the concerns of clients at the high level of inefficiency and waste and equally strongly reinforced the earlier message of the need for integration. The Egan Report differed from earlier reports by urging the importation of best practice in supply chain management from other sectors. The report stated:

> We are proposing a radical change in the way we build. We wish to see, within five years, the construction industry deliver its products to its customers in the same way as the best customer-led manufacturing and service industries.

The work by the client bodies since 1994 on the nature of the cultural changes, coupled with a major review of construction procurement by the National Audit Office, led to the publication of three definitive and matching best practice standards. In October 2000, the Prime Minister, Tony Blair, launched *Better Public Buildings*, a best practice standard he had commissioned jointly from the Commission for Architecture in the Built Environment (CABE) and the Treasury. In September 2000, the Confederation of Construction Clients launched their *Charter Handbook* and, in January 2001, the National Audit Office published *Modernising Construction*. The latter provides a very detailed appraisal of the true effectiveness of the industry and the barriers that inhibit reform. It also sets out the fully integrated approach that is essential if radically better performance and value are to be achieved and proven by measurement.

All three best practice standards used the lessons learned on total supply chain integration by the Building Down Barriers project, that had been launched in 1997 by the Ministry of Defence to adapt best practice in supply chain management from the manufacturing sector. The objective had been to develop a comprehensive set of tools to integrate and manage the design and construction supply chain that had been tested, refined and proved on two identical test-bed pilot projects. This had been jointly funded by Defence Estates and the Department of the Environment, Transport and the Regions and had led to the production of the unique *Building Down Barriers Handbook of Supply Chain Management*. An overview of the full, and very comprehensive, handbook toolset is available in *The Essentials*, which was published by *CIRIA* in July 2000. The toolset offers a systematic and managed approach to the procurement and maintenance of constructed facilities, based on integrating all the activities of a pre-assembled supply chain under the control of a single point of responsibility. The overall goal is to harness the full potential of the supply chain to deliver optimal value to the client, in terms of the through life performance of the facility, whilst improving the profits earned by all involved.

As a consequence of their use of a common foundation, the three best practice standards are completely attuned in their approach to: integration, whole-life costs, and value for money. The key objectives of the approach are to deliver the following:

- The finished building will ensure maximum functionality.
- The end users will benefit from the lowest cost of ownership.
- Inefficiency and waste in the utilisation of labour and materials will be eliminated.
- The specialist suppliers will be involved in design from the outset to achieve integration and buildability.
- The design and construction of the building will be achieved through a single point of contact for the most effective co-ordination and clarity of responsibility.
- Current performance and improvement achievements are to be established by measurement.

2 The international demand for improvement

The demand from clients in the UK for radical improvements in the performance of the construction industry is not unique; it is a demand that is echoed by clients across most of the developed world. This is exemplified particularly well in two documents that have been published since the Latham Report in 1994. The first was the 1995 report by the Construction and Building Sub-committee of the Committee on Civilian Industrial Technology (CCIT) in the United States of America. The CCIT is part of the National Science and Technology Council (NSTC), which is a cabinet-level group charged with setting Federal technology policy. The second document was Construction 21, which was produced for the Singapore Ministry of Manpower in 1999.

The 1995 Construction and Building Sub-committee Report

The focus of the report was the effect the performance of the construction industry had on the performance of other sectors and on their ability to be competitive. It went on to demand that the research and development sector of the construction industry adopt two priority thrusts, namely the development of technologies and practices which would deliver better constructed facilities, and the development of technologies and practices which would improve the health and safety of the construction workforce. It set a requirement that the technologies and practices had to be developed for use by 2003 and would deliver specific improvements against the 1995 baseline performance. The improvements became the 'National Construction Goals' and were as follows:

- 50% reduction in delivery time
- 50% reduction in operation, maintenance and energy costs
- 30% increase in productivity and comfort (of the occupants)
- 50% fewer occupant related illnesses and injuries
- 50% less waste and pollution
- 50% more durability and flexibility
- 50% reduction in construction work illnesses and injuries

The report pointed out that the construction sector constituted US$850 billion, which was about 13 per cent of GDP, and that the quality of constructed facilities was vital to the competitiveness of all US industry. It emphasised the need for a whole-life viewpoint of construction to give realistic attention to values and costs of constructed facilities. In support of this need, it cited the example of an office building, where the annual operating costs (including the salaries of the occupants) roughly equalled the initial construction cost. This meant that the primary value came from the productivity of the occupants, which depended on the capability of the building to meet user needs throughout its useful life.

The full text of the seven 'National Construction Goals' for better constructed facilities, and better health and safety of the workforce, is as follows:

- **50% reduction in delivery time.** Reduction in the time from the decision to construct a new facility to its readiness for service is vital to industrial competitiveness and to project cost reduction. During the initial programming, design, procurement, construction and commissioning process, the need of the client for the facility is not being met; needs to evolve over time so a facility long in delivery may be uncompetitive when it is finished; and the investments in producing the facility cannot be recouped until the facility is operational. The need for reduction in time to project completion is often stronger in the case of renovations and repairs of existing facilities because of interruption of ongoing business. Owners, users, designers and constructors are among the groups calling for technologies and practices reducing delivery time.

- **50% reduction in operation, maintenance and energy costs.** Operation and maintenance costs over the life of the facility usually exceed its first cost and may do so on an annualised cost basis. To the extent that prices for energy, water, sewerage, waste, communications, taxes, insurance, fire safety, plant services, etc., represent costs to society in terms of resource consumption, operation and maintenance costs also reflect the environmental qualities of the constructed facility. Therefore, reductions in operation and maintenance and energy costs benefit the general public as well as the owners and users of the facility.

- **30% increase in productivity and comfort.** Industry and government studies have shown that the annual salary costs of the occupants of a commercial or institutional building are of the same order of magnitude as the capital cost of the building. Indeed, the purpose of the building is to shelter and support the activities of its occupants. Improvement of the productivity of the occupants (or for an industrial facility, improvement of the productivity of the process housed by the facility) is the most important performance characteristic for most constructed facilities.

- **50% fewer occupant related illness and injuries.** Buildings are intended to shelter and support human activities, yet the environment and performance of buildings can contribute to illness and injuries for building users. Examples are avoidable injuries caused by fire or natural hazards, slips and falls, legionnaires' disease from airborne bacteria, often associated with workplace environment (sick building symptoms) and building damage or collapse from fire, earthquakes, or extreme winds. Sick building symptoms include irritation of eyes, nose and skin, headache and fatigue. If improvements in the quality of the indoor environment reduce days of productive work lost to sick days and impaired productivity, annual nationwide savings could reach billions of dollars. Criminal violence in buildings is a safety issue which can be addressed in part by building design. Reductions in illness and injuries will increase user's productivity as well as reducing costs of medical care and litigation.

- **50% less waste and pollution.** Improvements of the performance of constructed facilities that shelter and support most human activities, provide major opportunities to reduce waste and pollution at every step of the delivery process, from raw material extraction to final demolition and recycling of the shelter and its contents. Examples are reduced energy use and greenhouse gas emissions and reduced water consumption and waste water production. Waste and pollution also can be reduced in the construction process: construction wastes are estimated at 20–30% of the volume of landfills.

- **50% more durability and flexibility.** Durability denotes the capability of the constructed facility to continue (given appropriate maintenance) its initial performance over the intended service life, and flexibility denotes the capability to adapt the constructed facility to changes in use or users' needs. High durability and flexibility contribute strongly to the life cycle quality of constructed facilities since they usually endure for many decades.

- **50% reduction in construction work illnesses and injuries.** A factor affecting international competitiveness is the cost of injuries and diseases among construction workers. Although the construction workforce represents about 6% of the Nation's workforce, it is estimated that the construction industry pays for about one-third of the Nation's workers' compensation. Workers' compensation insurance premiums range from 7% to 100% of payroll in the construction industry. Construction workers die as a result of work-related trauma at a rate that is 2.5 times the annual rate for workers in all other industry sectors (13.6 deaths per 100,000 construction workers, as compared to 5.5

deaths per 100,000 workers in all other industry sectors). Construction workers also experience a higher incidence of non-fatal injuries than workers in other industries.

It is interesting to note the similarity between the non-technical barriers listed in the report and those listed in the Latham and Egan Reports in the UK in 1994 and 1998. The barriers listed in the USA report were as follows:

- lack of leadership
- adversarial relationships
- parochialism
- fragmentation of the industry
- inadequate owner involvement
- increasing scarcity of skilled labour
- liability

The 1999 *Construction 21* Report

As in the example of the USA report, the *Construction 21* Report took a long, hard and unbiased look at the weaknesses in the performance of the construction industry in Singapore and examined their causes. It compared the situation in Singapore with that in Australia, Japan, Hong Kong, the UK and the USA, and it examined developments elsewhere, such as in the Netherlands and in Denmark. It argued that the current state of affairs in the construction industry could not be sustained and that the industry must align its performance and practices with the other sectors of the economy. The Report compared the size of the Singapore construction industry with other countries and noted the similarity, in that it was 9.1 per cent of GDP in Singapore, 6.3 per cent in Australia, 10.4 per cent in Japan, 9.7 per cent in Hong Kong and 8 per cent in the UK.

The primary cause of the industry's performance weaknesses was seen as the segregation of design and construction, which formed a barrier to the consideration of buildability, savings in labour usage, ease of maintenance and safety at the design stage.

This segregation reduced the efficiency of the industry and led to much rework and wastage downstream.

Construction 21 calls for paradigm shifts in the performance and perceptions of the Singapore construction industry, namely that it should:

- Change from being perceived as **dirty, demanding and dangerous** to being perceived as **professional, productive and progressive.**
- Become a **knowledge industry** which compares well with other technologically advanced industries.
- Adopt a **distributed manufacturing approach** where construction products can be manufactured off-site and brought together on-site for assembly.
- Adopt an **integrated approach** to design, construction and maintenance where there is close co-operation and collaboration between consultants, construction contractors and manufacturers which lead to the formation of synergistic partnerships.
- Deliver **cost competitiveness through higher productivity.**

It is obvious from these two reports (particularly the *Construction 21* Report, which looks at the situation in several other countries) that there is universal recognition that a radical improvement in the performance of the construction industry would directly boost the competitiveness of other sectors. The reason for this is very simple: the results of the construction industry's poor performance are unnecessarily high and badly controlled initial costs; unnecessarily high and badly predicted maintenance and running costs; and poor functionality which adversely affects the efficiency (and therefore the cost) of the activities housed inside the building or facility. All these unnecessary costs feed directly into the overheads of the business activity housed by the building or facility. Consequently the prices charged to the customers of that business activity (be it a manufactured item or a service) are higher than they need be and are therefore not as competitive as they could be.

Most firms in other sectors have striven to improve their competitiveness by the measurement and elimination of the unnecessary costs in their manufacturing or retail supply chain, which come from the inefficient utilisation of labour and materials. At the same time, they have also improved the quality of their product to more effectively meet the aspirations of their customers. The effectiveness with which they have achieved this means they now have the knowledge and skills to look closely at the procurement process that delivers their buildings and facilities and they are able to compare it with their manufacturing or retail processes. They can clearly see the inefficiency and waste that occur at all stages of the design and construction process and can equally clearly see that it is caused by fragmentation and adversarial attitudes in the design and construction supply chain. They are worried by the inability to factor accurately the cost of ownership of their buildings and facilities into their long-term business plans that comes from lacking on accurate prediction of the cost of ownership from the design and construction team.

Firms do not like the unforeseen risks of ownership, which come from an unforeseen need to replace expensive components during the service life of the building or facility. As this component replacement cost had not been anticipated in their long-term business plans, it invariably had to be paid out of their profit margin, or by raising their prices.

The reality is that highly efficient customers of the construction industry throughout the developed world are demanding that their performance is replicated by the construction industry. This was the primary demand of the 1998 Egan Report in the UK, which said:

> We wish to see, within the next five years, the construction industry deliver its products to its customers in the same way as the best consumer-led manufacturing and service industries.

Compare this with a similar statement in the *Construction 21* Report:

The construction industry must be aligned with the other sectors of the economy in its performance and practices.

Compare it also with the following statement from the USA report:

Innovation in the US construction industry is an essential component for America's economic prosperity and well being.

It is now imperative that construction industry firms recognise and acknowledge the validity and nature of the demands for radical reform from its end user clients. The construction industry can no longer get away with delivering constructed products to its clients that ignore the affect those products have on the competitive performance of the end user client. It can no longer get away with the unjustified claim that it is different from all other sectors because it has to construct buildings in the open air and therefore improvements in supply chain management cannot be imported from other sectors. The message from the end user clients is very clear: the construction industry must embrace radical change and it must do so with urgency.

3 The three UK best practice standards and their six goals

Despite the intended impact on the UK construction industry of the Latham Report in 1994 and the Egan Report in 1998, there was growing concern by government, the external auditors of public sector procurement and a small number of leading edge repeat clients, that the traditional barriers to reform were proving unassailable. It was recognised that the primary reason for this was that the clients (particularly the internal professional advisers within their procurement groups) were refusing to change their traditional, sequential procurement practices and were unable to recognise that this was the main cause of the fragmentation and poor performance of the industry. This led to three concomitant, but independent moves to re-energise the reform by the publication of three best practice standards for construction procurement that could be imposed on clients by various external means. The three organisations that decided to take this proactive and courageous action were the National Audit Office, the Confederation of Construction Clients and the Cabinet Office (operating through the Department of Culture, Media and Sport, who worked with the Commission for Architecture in the Built Environment and the Treasury). The three best practice standards they developed were published in late 2000 and early 2001.

The Prime Minister became involved with the best practice standard commissioned by the Cabinet Office and imposed the radical changes in procurement practice, set out in *Better Public Buildings*, on the public sector when he launched it at 10 Downing

Street in October 2000. The Confederation of Construction Clients' *Clients' Charter* was launched in December 2000 and independent validation of compliance with the best practice standard set out in the *Charter Handbook* is a condition of attaining the status of a chartered client. The National Audit Office published its report in January 2001 and it intends to conduct all future audits of central government procurement bodies against the best practice standard set out in *Modernising Construction*. Similarly, *Better Public Buildings* and *Modernising Construction* will heavily influence future audits of local government and health authorities by the Audit Commission.

The importance of the impact of *Modernising Construction* and *Better Public Buildings* on the UK public sector clients should not be underestimated. The UK public sector is traditionally responsible for 40 per cent of the total annual expenditure of the construction industry and every public sector client is subject to external audit. As a consequence, the pressure of the external auditors will force a change in the procurement practices of 40 per cent of the industry's clients, who will be required to embrace the best practice standard that is common to *Modernising Construction* and *Better Public Buildings* in order to ensure their procurement practices deliver best value for the tax payer.

These intense external pressures on clients will inevitably ensure that a radical reform of client procurement practice in the UK becomes irresistible. Consequently, there is an urgent need for all involved in construction procurement in the UK to understand the key requirements that are common to all three best practice standards. Whilst this understanding should come from reading the three documents, this chapter of the guidebook endeavours to give busy practitioners in all sectors the priority areas for improvement.

Better Public Buildings

The report's primary thrust is targeted at the functional performance of the completed building and stipulates very firmly that well-designed buildings must enhance the quality of life for the end users. In his Foreword, the Prime Minister states:

The best-designed schools encourage children to learn. The best-designed hospitals help patients recover their spirits and their health.

The powerful secondary thrust in the report is the need to achieve better value over the whole lifetime of the building. Again, the Prime Minister states in his Foreword:

Integrating design and construction deliver better value for money as well as better buildings, particularly when attention is paid to the full costs of a building over its whole lifetime.

The report demands of public sector clients radical structural and cultural change in procurement practice, with the most fundamental and far-reaching being the requirement, in the 'Why and how' section, that their:

Procurement arrangements must enable specialist suppliers to contribute to design development from the outset.

This requirement should come as no surprise to anyone that has read and understood the 1998 Egan Report, *Rethinking Construction*, which very clearly saw total integration of design and construction and the use of supply chain management as key to better value for the end user client.

The report concludes by listing those actions that must stop and those that must be started. The key actions are as follows:

Stop:

- regarding good design as an optional extra
- treating lowest cost as best value
- valuing initial capital cost as more important than whole-life cost
- imagining that effectiveness and efficiency are divorced from design

Start:

- measuring efficiency and waste in construction
- appointing integrated teams focusing on the whole-life impact and performance of a development
- encouraging longer-term relationships with integrated project teams as part of long-term programmes, always subject to rigorous performance review
- using whole-life costing in the value-for-money assessment of buildings

Charter Handbook

The *Handbook* closely follows the theme of *Better Public Buildings* and sets out the obligations that define a best practice client in the 'Rethinking Construction' era. The purpose of the *Handbook* is to set out and describe a best practice standard to which Charter Clients must commit themselves. This is illustrated in the 'Background to the Charter' section of the *Handbook*, where the Construction Minister, Nick Raynsford, is recorded as saying that the Charter must set out:

> The minimum standards they (the clients) expect in construction procurement today, their aspirations for the future and a programme of steadily more demanding targets that will drive standards up in the future.

The *Handbook* recognises that for the current reform of the construction industry to succeed, it is imperative that the clients provide leadership for the essential and radical changes to the structure and culture of the entire supply chain through the reform of their procurement process. The *Handbook* requires Chartered Clients to lead the drive for continuous improvement of cultural relationships throughout the supply chain and of the constructed outputs of the industry, using performance measurement to provide proof of improvement.

The *Handbook* lists the obligations of a Charter Client, key to which are the following:

- Prepare a programme of cultural change with targets for its achievement over a period of at least three years' duration, but preferably five years or more.
- Measure their own performance in achieving their cultural change programme.
- Monitor the effects of implementing their programmes of cultural change, by calculating the national Key Performance Indicators that apply to their projects.
- Annually review and amend as necessary their cultural change programme in the light of what has been achieved.

The *Handbook* requires clients to have procurement processes that deliver (using measurement as the basis of proof) the following key improvements to the constructed products of the industry:

- major reductions in whole-life costs
- substantial improvements in functional efficiency
- a quality environment for end users
- reduced construction time
- improved predictability on budgets and time
- reduced defects on hand-over and during use
- elimination of inefficiency and waste in the design and construction process

The *Handbook* makes clear that best practice clients should always procure buildings and constructed facilities through integrated design and construction teams, preferably in long-term relationships, which involve all supply chain parties in the design process. It also requires the client to enforce the reforms in the structural relationships, culture, process and outputs of the supply-side by making them a condition of any relationship with the construction industry. Importantly, the *Handbook* also makes clear that consultants (especially architects) must be an intrinsic part of both the industry and the integrated supply chain.

Again, these requirements should come as no surprise to anyone that has read and understood the 1998 Egan Report, *Rethinking Construction*, which very clearly saw total integration of design and construction and the use of supply chain management as key to better value for the end user client.

Modernising Construction

Although this is the most detailed, comprehensive and specific of the three best practice standards, the *Modernising Construction* theme fully accords with that in *Better Public Buildings* and the *Charter Handbook*. The document is highly critical of the poor performance of the industry and the consequence this has for the public purse. It states:

> In 1999, a benchmarking study of 66 central government departments' construction projects with a total value of £500 million showed that three-quarters of the projects exceeded their budgets by up to 50% and two thirds had exceeded their original completion date by 63%.

The document lists and describes the major barriers to improved performance of the construction industry. The key barriers being:

- appointing designers separately from the rest of the team
- little integration of design teams or of the integration of the design and construction process
- insufficient weight given to users' needs and fitness for purpose of the construction
- use by client of prescriptive specifications, which stifle innovation and restrict the scope for value for money
- design often adds to the inefficiency of the construction process
- limited use of value management
- resistance to the integration of the supply chain
- limited understanding of the true cost of construction components and processes

- limited project management skills with a stronger emphasis on crisis management
- processes are such that specialist contractors and suppliers cannot contribute their experience and knowledge to designs
- insufficient weight given to users' needs and fitness for purpose of the construction

The document poses a series of key questions that public sector procurers need to consider if they are serious about improving quality and value for money. The most significant and radical of the key questions are as follows:

- Is supply chain integration achieved from the outset of the design process?
- Has the whole design and construction team been assembled before the design is well developed?
- What are the likely whole-life (running, maintenance and other support) costs?
- Have appropriate techniques been used, such as value management and value engineering, to determine whether the potential for waste and inefficiency has been minimised in the method of construction?
- Have efficiency improvements, to be delivered by the construction process, been quantified?

Finally, the document describes those areas where measurement of construction performance is essential. The priority areas being to measure:

- the cost effectiveness of the construction process such as labour productivity on site, extent of wasted materials, and the amount of construction work that has to be redone
- the quality of the completed construction and whether it is truly fit for the purpose designed
- the operational usage of completed buildings to determine whether the efficiency improvements that the original design was intended to deliver were achieved

Once again, these key questions should come as no surprise to anyone that has read and understood the 1998 Egan Report which very clearly saw total integration of design and construction and the use of supply chain management as key to better value for the end user client. Similarly, the priority areas where measurement is essential should come as no surprise to anyone that has read and understood the 1994 Latham Report, *Constructing the Team*, which focused strongly on the need to integrate the design and construction teams in order to eliminate the high levels of inefficiency and waste in the utilisation of labour and materials.

It is obvious from detailed analysis of the three standards that there are two key differences and six primary goals of construction best practice that mark out best practice procurement from all other forms of traditional procurement and are common to all three standards. The two key differentiators of construction best practice are:

1 **Abandonment of lowest capital cost as the value comparator.** This is replaced in the selection process with whole-life cost and functional performance as the value for money comparators. This means that industry must predict and be measured by its ability to deliver maximum durability and functionality (which includes delighted end users).

2 **Involving specialist contractors and suppliers in design from the outset.** This means abandoning all forms of traditional procurement that delay the appointment of the specialist suppliers (sub-contractors, specialist contractors and manufacturers) until the design is well advanced (most of the buildability problems on site are created in the first 20 per cent of the design process). Traditional forms of sequential appointment are replaced with a requirement to appoint a totally integrated design and construction supply chain from the outset. This is only possible if the appointment of the integrated supply chain is through a single point of contact – precisely as it would be in the purchase of every other product from every other sector.

The six primary goals of construction best practice are:

1   The finished building will deliver maximum functionality, which includes delighted end users.
2   End users will benefit from the lowest optimum cost of ownership.
3   Inefficiency and waste in the utilisation of labour and materials will be eliminated.
4   Specialist suppliers will be involved in design from the outset to achieve integration and buildability.
5   Design and construction of the building will be achieved through a single point of contact for the most effective co-ordination and clarity of responsibility.
6   Current performance and improvement achievements will be established by measurement.

4 The virtual firm

Any drive for improvement in any activity must start by knowing with absolute certainty where you are with your current performance and where you wish to arrive with your improved performance.

The imperative of measuring and understanding your current performance applies particularly to the utilisation of labour in each sector, by the design consultants, construction contractors, specialist suppliers (sub-contractors) and manufacturers. This will almost certainly be the weakest area of performance, and will be where the greatest potential for improvement and cost saving lies. The 1994 Latham Report, and subsequent work in the UK by the Building Research Establishment (BRE) CALIBRE team and the Building Services Research and Information Association (BSRIA), suggests that up to 30 per cent of the capital cost of construction is consumed by unnecessary costs and much of this is caused by the inefficient utilisation of labour throughout the design and construction supply chain, with the remainder caused by wastage of materials due to reworking.

Setting goals

Unless you know precisely how good you are, it is impossible to set sensible targets for improvement. If you are close to best-in-class, you only need to improve sufficiently to stay best-in-class. If you are well behind the best-in-class, you need to set tough targets for improvement over a short time span. You also need to establish what the precise current performance of your competitors is, and you must do so using the same form of measurement, so that you are comparing like with like.

It is also imperative to set goals for your improvement process that will deliver a constructed product that the end user customer wants to buy. You need to be absolutely certain that your goals relate precisely to the expectations of the true end user customer and not to assumptions made by others that may turn out to be false. What quality, cost and performance do the end users expect from the built product and how does the quality, cost and performance of your built products compare with that of the best-in-class? Only when you have established this firm baseline, by measuring your current performance and establishing the expectations of the end user, can you develop your improvement programme, set your improvement targets and devise the Key Performance Indicators to check your effectiveness at meeting the improvement targets you have set. To use the very simple example of a 1,500-metre runner, the first necessity is to know how fast he or she can run 1,500 metres. The next necessity is to know how fast the opposition can run the same distance, especially the best-in-class runners. Having established the gap in performance, the next requirement is to develop an improvement programme to get from where you are to where you want to be. It also helps if a knowledgeable coach or mentor can be employed to set a sensible and attainable improvement programme and to help check progress against the programme.

It follows from the above that it is imperative to establish goals for improvement that are targeted at those weaknesses in industry performance that end users believe have a negative effect on the value for money in the built products that they receive. Unless supply-side firms ensure their goals are the same as the demand-side end user's goals, the two sides will be heading in different directions. The publication of the three best practice standards (*Charter Handbook*, *Modernising Construction* and *Better Public Buildings*) now provides a route that can be used to establish the true expectations of the end users and thus avoid the risks inherent in making assumptions about what the end user expects.

Although reading the three publications is the ideal way of understanding the expectations of the end user, the previous chapters of this guidebook should simplify that process and thereby enable firms and organisations within each sector of the construction industry to recognise and prioritise those aspects of their working practices that are in most need of reform. In order to simplify the comparison even further, this chapter endeavours to set out briefly the key actions that are essential to the achievement of best practice.

At the end of the previous chapter it was made clear that there are two key differences that mark out best practice procurement from all other forms of traditional procurement. These are as follows:

- **Abandonment of lowest capital cost as the value comparator.** This is replaced in the selection process with whole-life cost and functional performance as the value for money comparators. This means that industry must predict and be measured by its ability to deliver maximum durability and functionality (which includes delighted end users).
- **Involving specialist contractors and suppliers in design from the outset.** This means abandoning all forms of traditional procurement which delay the appointment of the specialist suppliers (sub-contractors, specialist contractors and manufacturers) until the design is well advanced (most of the buildability problems on site are created in the first 20 per cent of the design process). Traditional forms of sequential appointment are replaced with a requirement to appoint a totally integrated design and construction supply chain from the outset. This is only possible if the appointment of the integrated supply chain is through a single point of contact – precisely as it would be in the purchase of every other product from every other sector.

The previous chapter also made clear that there are six goals of construction best practice and the six goals require every firm and every organisation involved in the design and construction supply chain to ensure that:

- Finished buildings will deliver maximum functionality, which includes delighted end users.
- End users will benefit from the lowest optimum cost of ownership.
- Inefficiency and waste in the utilisation of labour and materials will be eliminated.
- Specialist suppliers will be involved in design from the outset to achieve integration and buildability.
- Design and construction of the building will be achieved through a single point of contact for the most effective co-ordination and clarity of responsibility.
- Current performance and improvement achievements will be established by measurement.

The radically new form of construction procurement necessitated by the two key differentiators and the six primary goals of construction best practice, particularly the single point of contact, requires supply-side firms to embrace and understand the concept of a 'virtual firm'. This means that a group of firms, which constitute the entire supply chain for the design and construction of a typical building or constructed facility, must use long-term strategic supply chain partnerships to form themselves into a stable, mutually supportive supply-side alliance that works together and operates as a 'virtual firm'. The concept of a 'virtual firm' came out of the pioneering work on supply chain management done by the Building Down Barriers process development project, which was launched in 1997. This was intended to adapt best practice in supply chain management from manufacturing industry for use in the construction industry. The output from the project was the *Building Down Barriers Handbook of Supply Chain Management* (see Chapter 11, Further reading).

When the Building Down Barriers team had to explain how the long-term strategic supply chain partnerships that are the foundation of the Building Down Barriers approach worked, it seemed logical to describe the long-term relationship between the supply-side firms as a 'virtual' relationship since it did not require a formal contract or sub-contract, nor did it necessitate takeovers or mergers. Others have since undertaken to develop the concept of a 'virtual firm', such as the Design Build Foundation and Reading Construction Forum, and those wishing to gain from their development work should contact them.

The *Building Down Barriers Handbook of Supply Chain Management* says of long-term supplier relationships:

> Long-term relationships can drive up quality and drive down both capital and through-life costs for clients. At the same time, they can increase profitability for the supply chain. These long-term relationships are likely to be with only a small number of suppliers in each key supply category, because it is not possible to invest in the kind of relationship required with a large number of organisations.

Recognising and understanding the high level of commonality between apparently differing building types, when they are broken down into components, materials and processes, should help in the formation of these strategic supply-side partnerships. All too often, the cry is heard that 'every building is unique and different', yet when the building is broken down into components and materials a different picture emerges. Steel frames are remarkably common to offices, hospitals, health centres, warehouses, multiple-occupancy living accommodation, libraries, workshops, factories and hotels. Brickwork and blockwork occurs in every building type, from high rise tower blocks to housing. Windows are common to every building type, with the only real variation being the type of material. Electrical services are also common to all building types, with minor variations where there is a requirement for specialist components. Even mechanical services have a considerable commonality across all building types.

Benefits of standardisation

The reality of the benefits that can come from a greater commonality of components and materials across differing building types was picked up in the UK in the 1998 Egan Report, *Rethinking*

Construction. The report was highly critical of the UK construction industry's unwillingness to grasp the benefits of greater standardisation of components and materials across differing building types. It stated:

> We see a useful way of dealing efficiently with the complexity of construction, which is to make greater use of standardised components. We call on clients and designers to make much greater use of standardised components and measure the benefits of greater efficiency and quality that standardisation can deliver . . . Standardisation of process and components need not result in poor aesthetics or monotonous buildings. We have seen that, both in this country and abroad, the best architects are entirely capable of designing attractive buildings that use a high degree of standardisation.

The Egan Report also cited examples of a lack of standardisation of components in the UK, namely:

- Toilet pans: there are 150 different types in the UK, but only six in the USA.
- Lift cars: although standard products are available, designers almost invariably wish to customise these.
- Doors: hundreds of combinations of size, veneer and ironmongery exist.
- Manhole covers: local authorities have more than thirty different specifications for standard manhole covers.

Case history: Building Down Barriers

Evidence of what can happen when the specialist suppliers are linked closely with designers and construction contractors within long-term strategic supply chain relationships was clearly shown on the two buildings used to test the application of the supply chain management tools and techniques. These pilot projects achieved many outstanding improvements in performance and in outputs that came directly from the involvement of specialist suppliers in design from the outset. Not only were these outstanding improvements at project level, the specialist suppliers could see that if they continued to work together with the designers and construction contractors at strategic level, they could continue to improve their performance on a project-by-project basis. The steel fabrication firm on one of the pilot projects achieved major savings in the capital cost of the steel frame and a major improvement in their profit margin. In addition, they were convinced they could take 15 per cent off the capital cost of any subsequent steel frame if the design and construction team could stay together. The specialist suppliers on both pilot projects were convinced of the commercial benefits that could flow directly from enabling them to work with the consultant designers at a strategic level to eliminate the recurrent causes of disruption and abortive work, so that 'right first time' on site could be achieved every time for every project. At pilot project level, this involvement of specialist suppliers in design from the outset led to a far greater use of standard components and materials, which was not imposed by the supply chain management tools and techniques or by the architect or engineers, but came solely from the direct involvement of specialist suppliers and manufacturers at concept design stage. Examples of the improvements measured on the two buildings that came directly from this way of working together were: a 20 per cent reduction in construction time; wastage and rework levels consistently below 2 per cent; labour-time spent overall on adding value to the building in the region of 65–70 per cent; no reportable accidents; no claims; an absence of commercial or contractual conflict throughout the two supply chains and a high level of morale on site.

The 'virtual firm' always works together when dealing with a client whose procurement process embraces the six primary best practice goals, since such a client would want the same single point of contact and the efficient supply chain management that would be the norm when buying non-construction products. The client's point of contact with the 'virtual firm' should be that which makes best sense to the members of the 'virtual firm' and may not be the construction contractor, as it would be in traditional procurement.

For other clients, particularly small and occasional clients, the supply-side firms in the 'virtual firm' act as the situation dictates, either operating as single entities or in cluster groupings, whilst still giving the client the benefit of the improvements to the construction process they have developed within their long-term strategic partnerships. By working together in this mutually supportive way, each firm will be able to ensure and improve its profitability and its market share, no matter what procurement approach the client adopts.

The creation of a 'virtual firm' on the supply-side of the industry can be driven by a major repeat client that is determined to force the pace and direction of reform in order to achieve better value from construction procurement and thus reduce its impact on their overheads. There are several excellent examples of this in the UK with clients such as Argent, BAA, McDonald's and several other major retailers that have utilised their skill at managing their retail supply chains to manage their design and construction supply chains.

This passive supply-side response to demand-side pressure from individual clients for supply chain integration is neither the only nor the best way of driving forward the radical reforms that supply chain integration demands. This is because of the risk that supply-side firms may only integrate when working for that specific client and may continue to operate in an inefficient, fragmented and adversarial way for all other clients. The more effective way of introducing supply chain integration and the 'virtual firm' is where the initiative is taken by the supply-side and it becomes the way they do their business for all their clients. Where this occurs, supply chain integration has a far greater chance of being introduced because the firms in the 'virtual firm' are driven by a mutual recognition of the very real commercial benefits that can accrue to them all from the elimination of unnecessary costs and the delivery of better quality. This supply-side initiative tends to be led in the UK by major construction contractors who have recognised that the market is becoming more intelligent and discerning and those firms that act first to radically improve the performance of themselves and their supply chains stand the greatest chance of maintaining or improving their share of that more discerning market.

This concept of a 'virtual firm' may be more easily understood if compared to the operation of a major football team such as Manchester United or a major rugby team such as St Helens. The skills required from those selected for a given game will vary, depending on the make-up of the opposing team; and the more closely the skills of the home team can match or exceed those of the opposing team, the greater the likelihood of the home team winning. This necessitates the existence of a squad of players that exceeds those needed for a given game, so that the manager is able to choose a different team make-up for an individual game. A wise and successful manager will carefully analyse the techniques and skills of the players in the specific opposing team to better understand the precise requirements of the specific game. The wise and successful manager will also carefully study the performance track record of the specific opposing team to assess its strengths and its weaknesses. Having thus established the likely skills requirement for the specific game, the manager will compare those precise requirements with the skills, experiences and temperaments of the individual players in his full squad (which may well be double or treble the number of players needed for an individual game) and will pick a team, plus a small number of reserves, for the given game.

From this analogy it can be seen that the operation of the 'virtual firm' is closely akin to that of a major football or rugby team. The full squad of strategic supply chain partners will need to be a good deal more than would be required for any given project, because it must encompass the full range of skills and experience necessary for the wide variety of design and construction needs that will occur across the full range of building and civil engineering types the 'virtual firm' might wish to embrace. From this full squad, the 'virtual firm' selects the team that is appropriate for a given project. This will need to include a small team of reserves to cover situations that might arise during the design and construction process, i.e. a switch from steel frame to concrete frame, or an unforeseeable overload of an individual firm for reasons outside the given project.

Developing loyalty

There is another aspect of the management of a successful football or rugby team that has great relevance to the management of a successful 'virtual firm'. Consistently high performance of the team would be impossible if the players were constantly changed for every game, without the opportunity to practise together in the relatively stable, full squad. There would be no empathy or loyalty between the players, they would have virtually no understanding of each other's skills, experience and temperament, they would never have had the opportunity to regularly train or play together as a co-ordinated and mutually supportive team. The critically important need to develop team skills, loyalty and a common goal for the team (which must apply to the full squad as much as the team for a given game) would be impossible. The manager would find the selection of the most appropriate and effective team for a given game an endless and thankless task and would probably be driven to selecting the team for a given game with a pin! (or by use of lowest price tendering from specialist suppliers in the case of the construction industry).

The successful team requires the manager to have a full and detailed knowledge of every aspect of the skills and experience of the full squad. The manager must also have total confidence in the loyalty of every member of the full squad and must have total belief in their shared understanding of mutually agreed goals. With this firm foundation for the selection process, the manager can be reasonably certain that the team selected for any given game can be relied on to perform as an effective, efficient, enthusiastic and mutually supportive team.

These considerations are equally relevant to the effectiveness and success of the 'virtual firm'. All too often projects suffer because the design and construction team are cobbled together for the first time and have no expectations of ever being together in the future.

Worse still, most of them will have been selected on a lowest price basis, where profit margins have been squeezed to the bone and the only way of making a decent profit may well be through claims against other team members, or against the client. Even worse is the fact that many team members will be introduced long after the game has started, since most specialist suppliers (sub-contractors, specialist contractors or manufacturers) will not be appointed until the design is well advanced or even complete. Consequently, their skill, experience and knowledge will be ignored in the development of the design, even though harnessing it from the outset of design development may well have improved the cost effectiveness and buildability of the constructed solution and thus improved the profits of all concerned.

Any hope of creating a mutually supportive, loyal and highly skilled team in this environment is clearly an impossibility, and the resulting fragmentation and adversarial attitudes are the cause of the situation which is very accurately portrayed in the National Audit Office Report, *Modernising Construction*, with the statement:

> In 1999, a benchmarking study of 66 central government departments' construction projects with a total value of £500 million showed that three-quarters of the projects exceeded their budgets by up to 50% and two thirds had exceeded their original completion date by 63%.

If all of us can understand and appreciate what makes a successful team in top level football or rugby, why cannot we transfer that understanding to our own construction industry? In the case of the construction industry, 80 per cent of the team members are drawn from the specialist suppliers sector of the industry, and because they are not part of a 'virtual firm', they are generally selected on a project-by-project basis by the lowest price they can tender for the individual project. Consequently, their long-term security and profitability are high risk, the rate of bankruptcy is far higher than in other industries, the entry level is dangerously low and their valuable skill and experience is rarely ever harnessed to drive out unnecessary costs and drive up quality.

How often do we as individuals join the endless and constant chorus of complaints about 'cowboy builders' or 'cowboy suppliers' in our domestic lives, but operate in a manner that enables the existence of such firms when we switch to our corporate lives? We all need to rethink the design and construction process and replicate the experience of successful firms in other sectors. They have demonstrated quite clearly that their success is founded on long-term strategic supply chain partnerships that embody the seven principles of supply chain management described in the Building Down Barriers *Handbook of Supply Chain Management*. The 1998 Egan Report, *Rethinking Construction*, made very clear that integration of design and construction and the use of best practice in supply chain management must also be the foundation of successful 'virtual firms' in the construction industry. The Egan Report stated:

> We are proposing a radical change in the way we build. We wish to see, within five years, the construction industry deliver its products to its customers in the same way as the best customer-led manufacturing and service industries.

The seven universal principles of supply chain management are described in depth in the Building Down Barriers *Handbook of Supply Chain Management*, but are briefly as follows:

- **Compete through superior underlying value.** Key members of the supply chain work together to improve quality and durability, and to reduce underlying costs (labour and materials elements of the component and process costs) while improving profits. This is primarily about ending reworking and the achievement of a 'right first time' culture.
- **Define client values.** This requires all members of the supply chain (from end users to manufacturers) to work together, using formal value management techniques, to define and record the detailed business needs of the end users that must be delivered efficiently by the built solution. This ensures specialist suppliers have a detailed understanding of the end user's functional requirements.

- **Establish supplier relationships.** The products and services of the specialist suppliers account for over 80 per cent of the total cost of construction. It is therefore essential for the entire design and construction supply chain to establish better and more collaborative ways of working together, so that the skills throughout the supply chain can be harnessed and integrated to minimise waste of labour and materials. These better and more collaborative ways of working should also encourage exploitation of the latest innovations in equipment, materials and building processes.

- **Integrate project activities.** This involves breaking down the construction activities into sub-systems or clusters. These are relatively independent elements of the whole building or facility, such as groundworks, frame and envelope, mechanical and electrical services or internal finishes. Within each sub-system or cluster, the design, construction, material and component suppliers work together in close collaboration to develop detailed designs, construction methods and actual prices for delivery.

- **Manage costs collaboratively.** This involves 'target costing' where suppliers work backwards from the client's functional requirements and gross maximum price (cost budget). The supply chain, particularly in the cluster groupings, work together to design a product that matches the required level of quality and functionality and provides a viable level of profit for all at the agreed target price (which must be within the gross maximum price).

- **Develop continuous improvement.** The supply chain members openly measure and assess all aspects of their current performance, including the effective utilisation of labour and materials. They then agree continuous improvement targets for each aspect of design and construction activity that will deliver maximum savings in underlying process and materials costs.

- **Mobilise and develop people.** All involved must recognise that their staff will need to learn new ways of thinking, acting, and reacting. This involves 'unlearning' old ways and recognising the challenges to be met and the resistance and difficulties that can be expected.

At its simplest, strategic supply chain partnering (or lean construction) is the means by which the supply-side firms work together to drive out all forms of unnecessary costs and to drive up the quality of the constructed product. It is the foundation of every supply-side firm's ability to compete effectively for work in any situation. When we buy a manufactured product from any other sector (such as a television, a car, or a ship) we do not expect to have to enter into a partnership arrangement in order to ensure value for money. In most cases (as it is in the construction industry, where the majority of clients are small and occasional) the purchase will be one-off and any form of partnership between the client and the supply-side would be of very limited value in driving out unnecessary costs. There will, of course, be the occasion when a large number of identical or similar products will be required over a period of time by an individual client and this may well make a partnering arrangement sensible for a particular client in a particular instance. Nevertheless, the lesson from other sectors, and the message from the 1998 Egan Report, is that partnering will deliver the greatest improvements in performance where it is the basis of the long-term strategic relationships between firms on the design and construction supply-side of the industry.

This necessitates a radical and profound change in the way the supply-side firms operate and this in turn requires their Chief Executives to understand the nature of the changes in working practices that must be put in place within their own firm and within the firms with which they do business. The Chief Executives must also measure their organisation's current performance (such as the effective utilisation of labour and materials) and that of their suppliers so that they have a firm basis from which to start the improvement process. They must then become fervent champions for the changes in working practices, because only powerful and clear-sighted leadership from the Chief Executive can make those changes happen.

The magnitude of these changes should not be underestimated; they will affect everyone in the design and construction supply chain and they will not happen without a major change programme and the investment in carefully structured training and mentoring.

Measurement of performance will be difficult at first since it has been rarely done in the construction industry and the results may be hard to accept, both for the organisation and for the individual concerned. This will be especially so where it relates to the effective utilisation of labour and materials and the initial measurements validate the low levels assumed by the Latham Report and confirmed by the work of the Building Research Establishment CALIBRE team and Building Services Research and Information Association Technical Note 14/97 (see Chapter 11, Further reading). Some people will find the changes threatening and will endeavour to thwart them to maintain the *status quo*. Because of the heavy baggage they carry, of established and comfortable custom and practice, many people will find it difficult to understand the reason for the new ways of working and this will require the Chief Executive to ensure that the message is expressed in simple, easy to understand terms.

The lesson from those organisations that have successfully and radically improved their performance is absolutely clear. Any drive to radically improve performance will not be successful unless it is seen to be led by a Chief Executive who obviously understands the nature and magnitude of the changes in working practices and who can be seen to be clearly and implacably determined to make those changes happen.

5 Motivating and leading radical improvement

Previous chapters have described the external pressures that are forcing all sectors of the UK construction industry to embrace radical and profound change. They have also described the very real benefits that can accrue from an integrated supply-side in terms of higher and more assured profits, better whole-life value and lower prices. However, no matter how intense the external pressures for change, radical and profound changes within an individual firm will not occur unless a Chief Executive who clearly understands the nature and purpose of the changes very overtly and directly champions them. This applies whether the firm or organisation is on the client's demand-side or on the industry's supply-side.

It is also true that a major change in working practices will not happen without a great deal of concerted effort on the part of everyone in the organisation and an acceptance that all will have to change the way they work to some degree, with some having to change profoundly. Such a change can easily be thwarted by the inherent and powerful inertia of established custom and practices, and by covert resistance at all levels, especially at senior management level. It must be obvious to all that the change process is structured and directed by the Chief Executive in person, so that every action and every utterance of the Chief Executive constantly reinforces the direction and the urgency of the change. The experience of organisations that have successfully embraced radical and profound improvement teaches that the change process needs to be focused into four key areas if it is to be successful. Without all four being in place and operating concurrently, the change will become nothing more than wishful thinking and will soon be forgotten and replaced by the next bright idea.

The four essential and interlinked ingredients of successful change are as follows:

1   A clearly explained and rational goal that all can understand, with which all can identify, and which can be related to specific improvements in performance that can be measured and compared with current performance.

2   Committed, determined and overt leadership by the Chief Executive that leaves no one in any doubt about where the change must take the organisation, why it is commercially essential to get there, and the timescale for the change.

3   A detailed and comprehensive action plan for the development and implementation of the changes in working practices which explains in simple, easy to understand language what must be done differently by every member of the organisation. Adequate and appropriate training must support this for those that are required to operate in a different manner.

4   A simple and easy to understand explanation of the commercial benefits that will be delivered by the changes in working practices. This is best expressed in terms that relate to improved product quality, improved process efficiency, reduced waste in the production process and, most importantly, reduced costs and increased profits.

In any organisation, in any business sector, major changes in working practices are extremely difficult to initiate and achieve. Evidence shows that it is rare if as many as 30 per cent of the workforce is in favour of the changes where they affect their own working practices. Another 30 per cent will fight against the changes (usually in covert ways which will be concealed from the Chief Executive) because they are afraid that the changes will adversely affect their status, their ability to perform, or their pay. The remainder will sit on the fence until they are convinced the changes are inevitable and beneficial.

Those covertly against the changes in working practices are most likely to be at middle and senior management level, including Board level. They are generally older and inevitably carry more baggage, in the form of being more wedded to the familiar, comfortable and trusted ways of working. Quite often, they are also not unsurprisingly worried that they are going to find it difficult to learn the new and unfamiliar ways of working at their time of life, and that this is likely to mean that their position in the hierarchy of the organisation will suffer as the younger and more junior staff learn more quickly the new ways of working and see their superiors struggling to cope.

Case history: Aerospace industry

The above can be illustrated by an actual example of a major international firm from the aerospace industry that achieved a dramatic improvement in performance in a remarkably short space of time. The Chief Executive likened his role to that of a crusader king. He said he had to be constantly seen, by every one of his troops, to be leading the way forward into battle. Every move he made, every phrase he spoke and every word he wrote had to reinforce and clarify the changes in working practices he wanted the firm to make. He had to ensure that everyone in the firm (and he said this must literally include every last person employed in his firm) must understand where the firm was going, why it must go there, what would happen if it failed to reach its destination, and (most importantly) what each individual had to do differently as their part of the change process. He said it was imperative that the tea lady and the cleaners felt they were included in the change process. They must understand why the changes in working practices were commercially essential and they must want to be an active part of the change process.

He also said that it was important to recognise and reward those that were making the greatest contribution to the change process. Quite often, the reward need be no more than public recognition by the Chief Executive for their efforts (i.e. a personal letter from the Chief Executive which is also put into the firm's newsletter). It was essential to provide everyone with regular progress reports which explained how the improvements in performance were being measured and what was being achieved in the various parts of the firm. Equally important, was the need to expose and deal with those that were blocking and opposing the changes in working practices. He said you could be absolutely certain that the grapevine would ensure that everyone would be aware of the names of those that were trying to block the changes. It was equally certain that everyone was watching to see if the Chief Executive was on the ball and would pick up on what was really happening. If the blockers were ignored, the message the grapevine would take around the firm was that the Chief Executive was not serious about the changes in working practices, so they could be safely ignored.

It can be seen from the above example that two important ingredients of a successful change process are communication and education. It is totally unreasonable to expect people to embrace radical change unless it has been explained in a language they can understand, and unless they have been trained in the new working practices. It is also not acceptable to assume the explanation is adequate without verifying if it has been understood, and the responsibility for selecting the most appropriate language is that of the sender of the message. It is imperative to validate whether the explanation has been fully understood at all levels and whether the training has actually changed working practices, preferably by the use of an objective, bottom-up feedback mechanism that has a good track record of success. It should be borne in mind that it has always been a reality of buying and selling that you can generally buy using your own language, but you can rarely sell unless you use the language of the potential buyer. Consequently, those 'selling' the message about radical change within an organisation need to use the language of those that need to 'buy' the message and this may require the message to be differently phrased for different recipients.

Measuring improvement

An excellent tool to measure and test the rate of improvement and to provide well-structured support for the change process is available in the UK in the form of the European Foundation for Quality Management (EFQM) Business Excellence Model. This has an outstanding track record across the private and the public sectors, and in large as well as small organisations in the UK and across Europe. The 'Award Simulation' self-assessment system provides an excellent mechanism for providing a regular, annual, consistent and very objective measurement of the rate of improvement from the people at the sharp end of the organisation. It tells the Chief Executive and the Board what is really happening at the workface, and it provides the workforce with an opportunity of ensuring that the truth about what is going wrong in current working practices will get to the Chief Executive and the Board without being filtered or massaged by middle and senior managers.

It also forces organisations to adopt a very structured approach to improvement by the use of nine inter-dependent criteria, namely:

1   **Leadership.** How leaders develop and facilitate the achievement of the mission and vision, develop values for long-term success and implement these via appropriate actions and behaviours, and are personally involved in ensuring that the organisation's management system is developed and implemented.
2   **People.** How the organisation manages, develops and releases the knowledge and full potential of its people at an individual, team-based and organisation-wide level, and plans these activities in order to support its policy and strategy and the effective operation of its processes.
3   **Policy and Strategy.** How the organisation implements its mission and vision via a clear stakeholder focused strategy, supported by relevant policies, plans, objectives, targets and processes.
4   **Partnerships and Resources.** How the organisation plans and manages its external partnerships and internal resources in

order to support its policy and strategy and the effective operation of its processes.

5 **Processes.** How the organisation designs, manages, and improves its processes in order to support its policy and strategy and fully satisfy, and generate increasing value for, its customers and other stakeholders.

6 **People Results.** What the organisation is achieving in relation to its people.

7 **Customer Results.** What the organisation is achieving in relation to its external customers.

8 **Society Results.** What the organisation is achieving in relation to local, national and international society as appropriate.

9 **Key Performance Results.** What the organisation is achieving in relation to its planned performance.

Of the nine EFQM Business Excellence Model criteria, the most important are Leadership, Processes, Customer Results and Key Performance Results, and this is reflected in the marking system that gives the following shares of the total marks available:

- Leadership 10 per cent
- Processes 14 per cent
- Customer Results 20 per cent
- Key Performance Results 15 per cent

At the start of this chapter four key essentials of any successful change process were listed. These were as follows:

1 A clearly defined goal for the change process.
2 Committed leadership from the Chief Executive.
3 Well-defined and clearly described processes for improved working practices.
4 A clear explanation of the commercial benefits the change process will deliver.

Not surprisingly, these four essential ingredients have a close correlation with the four most important criteria from the EFQM

Business Excellence Model. Leadership is about setting a clear and unambiguous goal for the organisation and communicating that goal to everyone in a language each can understand. Process is about providing everyone with a route map that enables each of them to modify or change their out-moded working practices for best practice in a consistent way across the entire organisation. Customer Results and Key Performance Results are about measuring the commercial benefits that are delivered by the changes in working practices.

The overwhelming evidence from those organisations in every sector of private and public activity that have successfully embraced radical change is absolutely clear. No matter how much superficial enthusiasm there is for radical change, it will be stillborn unless the Chief Executive is seen to be totally committed to the change. This commitment must include the Chief Executive giving a very simple and straightforward explanation of the goal for the change process that must be expressed in terms that everyone (without exception) can understand, and there must be some way of testing the understanding at an individual level across the organisation. Making assumptions about what people understand can be very misleading and can lead to everyone going off in different directions because their interpretation of the message has been slightly different. It is imperative that their understanding is tested and validated, and the message must be rephrased if it is discovered that it has been interpreted differently across the organisation.

The change will also be stillborn if it is unwisely assumed that everyone will be able to modify his or her entrenched traditional working practices without considerable and appropriate help and education. The education needs to be carefully crafted and structured to address the areas where the working practices of the organisation do not match the best practice of the three standards (*Modernising Construction*, *Charter Handbook* and *Better Public Buildings*) and later chapters advise on a questionnaire that may assist in assessing where weaknesses lie and education needs to be targeted. The Construction Industry Training Board in the UK is liaising with various organisations to develop appropriate training workshops and courses. A good example of this is the ICOM/CITB link-up which offers a Diploma course in 'Construction Process Management' that uses the Construction Best Practice Programme booklet, *A Guide to Best Practice in Construction Procurement* (see Chapter 11, Further reading, for details).

The inertia of the deeply embedded traditional working practices will inevitably be powerful enough to neutralise any change process, no matter how apparently beneficial it appears to the Chief Executive. Consequently, the Chief Executive will need to appoint a team to develop the new working practices and also to work with the staff and operatives to ensure the application of those new working practices throughout the organisation. To ensure the maximum effectiveness of the team, it may need to work directly with the Chief Executive so that everyone in the organisation is left in no doubt about the Chief Executive's intentions and there is no possibility of the team being side-lined by those senior staff that are covertly against change.

Of paramount importance to the successful adoption of the new working practices will be the provision of appropriate training and coaching in their appreciation and application. It is unreasonable and absurd to assume that people who have always worked in one way can suddenly work in a radically new way without a good deal of help and support. This must take the form of a detailed route map for the process that constitutes the new working practices and training in its use by trainers that fully understand the new processes, their purpose, the background behind the new processes and the products they must deliver to the customers (internal and external customers).

Finally, the Chief Executive must ensure there is some way of measuring the improvements in performance that are being delivered by the new working practices. This also needs to be communicated in simple, straightforward terms to everyone in the organisation at reasonably frequent intervals. It has always been said that 'success breeds success' and this is equally true of organisational change. As long as those involved in operating the changed working practices can see real and tangible improvements in the products those new working practices are delivering to their customers (and these can be internal as well as external customers) they will be motivated and enthused to continue to do things in this new way. Consequently there will be far less chance of the inertia of traditional working practices dragging them back to the old and inefficient ways of working.

Case history: Building Down Barriers

Whilst the morale of those at the sharp end will be boosted when they are an intrinsic part of the change process that delivers improved products to the client, the converse is also true and their morale can be devastated if the changed working practices are not sustained. In the case of a steel fabrication firm involved on one of the two Building Down Barriers pilot projects, the morale of the fabricators and erectors was sky high at the completion of the Building Down Barriers project. Their skill and experience had been harnessed from the outset of the design of the steel frame and they had therefore been very much instrumental in achieving 'right first time' in the fabrication and erection of the steel frame. Unfortunately, the subsequent project was a conventionally procured warehouse and their skill and experience were not utilised by the consultant designers despite intense lobbying by their Chief Executive and warnings from him that the design was flawed. The outcome was a steel frame where the design was riddled with the usual errors and the fabricators knew that parts of the steel frame they were fabricating would have to be taken down after erection and would have to come back to be modified. The erectors also knew that they would have to dismantle parts of the frame after erection, take them back to the factory for modification, and then return to site to re-erect them. Needless to say, their Chief Executive said their morale, that had been lifted to unprecedented heights by their Building Down Barriers experience, was devastated by their subsequent warehouse experience. Yet their Chief Executive said the warehouse job was quite normal and the flaws in the design of the steel frame were no worse than usual, the problem was that the erectors and fabricators had assumed that what happened on the Building Down Barriers project was so sensible and beneficial it would be replicated on all subsequent projects for all other clients.

In conclusion:

- unless the Chief Executive is prepared to be seen by every member of the workforce to be totally committed to the change programme
- unless the Chief Executive is prepared to explain the goal for the change programme in simple and easy to understand language
- unless the Chief Executive is prepared to ensure the new tools (working practices) are developed and applied
- unless the Chief Executive is prepared to ensure appropriate training is available and that staff are able to attend it
- unless the Chief Executive is prepared to ensure the improvements in performance are regularly measured and the results published to the workforce
- unless the Chief Executive believes the changes should be the way the organisation operates for all its customers

then there is little point in even thinking about introducing a radical change in the way the organisation operates.

This all-important leadership role of the Chief Executive applies to every firm or organisation in the design and construction supply chain. It applies equally to the end user as it does to the manufacturer, since the application of the two key differentiators and the six primary goals of construction best practice from the three best practice standards require radical changes in the way all of them operate.

This all-important leadership role also applies to the vast number of institutes and trade organisations within the UK construction industry, which must become an intrinsic and supportive part of the change process if the best practice approach of the three standards is to take hold and flourish. Their Chief Executives must provide co-ordinated leadership for the new, integrated way of designing and constructing which delivers the six primary goals of best practice. Unless they do so, the traditional fragmentation and confusion will continue, where each institute and each trade organisation appears to be heading in a different direction with a different view of what constitutes best practice.

The further education establishments must also play their part in supporting the change process. They must ensure that their vision for the culture, structure and working practices of the industry matches that of the three standards (*Modernising Construction*, *Charter Handbook* and *Better Public Buildings*) and their six goals of construction best practice. They must also ensure the graduates they produce share a vision of the reformed industry that accords with the direction those at the leading edge of reform have taken and with the true expectations of the end user clients. They must abandon any educational assumptions they have that might continue to reinforce the traditional fragmentation and adversarial attitudes of the industry, and they must aid the direction and the pace of reform towards the goal of total integration of design and construction and the delivery of best value.

6 Becoming a best practice client

Earlier chapters have explained the consequences for the UK construction industry of the publication of the Confederation of Construction Clients, *Charter Handbook*, the National Audit Office Report, *Modernising Construction*, and the Department of Culture, Media and Sport Report, *Better Public Buildings*.

The existence of these three publications, particularly *Modernising Construction* and *Better Public Buildings*, means that all public sector clients (who represent 40 per cent of the total UK construction market) will inevitably have little option but to adopt the best practice approach described in the National Audit Office and Department of Culture, Media and Sport reports. They are all subject to external audit by the National Audit Office (for central government procurers of construction services) or by the Audit Commission (health authority and local government procurers of construction services) and as these two audit bodies will inevitably use *Modernising Construction* and *Better Public Buildings* as the common evaluative criteria for all their assessments, those audited by them will have little option but to adopt the same approach to best practice. The private sector will obviously have far greater freedom to choose an appropriate approach to procurement. However, if they choose to become a Chartered Client within the Confederation of Construction Clients organisation, they will have to submit their procurement practices for evaluation against the Confederation of Construction Clients' *Charter Handbook* and will not be accepted as a Chartered Client unless the two match.

In earlier chapters it was made clear that the *Charter Handbook*, *Modernising Construction* and *Better Public Buildings* are extremely similar and throw up the same two key differentiators and the same six primary goals of construction best practice. The two key differentiators of construction best practice are:

1   **Abandonment of lowest capital cost as the value comparator.** This is replaced in the selection process with whole-life cost and functional performance as the value for money comparators. This means industry must predict and be measured by its ability to deliver maximum durability and functionality (which includes delighted end users).

2   **Involving specialist contractors and suppliers in design from the outset.** This means abandoning all forms of traditional procurement which delay the appointment of the specialist suppliers (sub-contractors, specialist contractors and manufacturers) until the design is well advanced (most of the buildability problems on site are created in the first 20 per cent of the design process). Traditional forms of sequential appointment are replaced with a requirement to appoint a totally integrated design and construction supply chain from the outset. This is only possible if the appointment of the integrated supply chain is through a single point of contact – precisely as it would be in the purchase of every other product from every other sector.

The six primary goals of construction best practice are:

1   The finished building will ensure maximum functionality.
2   The end users will benefit from the lowest cost of ownership.
3   Inefficiency and waste in the utilisation of labour and materials will be eliminated.
4   The specialist suppliers will be involved in design from the outset to achieve integration and buildability.
5   The design and the construction of the building will be achieved through a single point of contact for the most effective co-ordination and clarity of responsibility.

6   Current performance and improvement achievements will be established by measurement.

In the previous chapter the actions required to motivate and lead radical change within any organisation and within any sector of the construction industry were explained and it was emphasised that it was of paramount importance for the Chief Executive to adopt a powerful crusading role if the change were to succeed. The Chief Executive must ensure the following four ingredients are in place, namely:

1   A clearly explained and rational goal that all can understand, with which all can identify, and which can be related to specific improvements in performance that can be measured and compared with current performance.

2   Committed, determined and overt leadership by the Chief Executive that leaves no one in any doubt about where the change must take the organisation, why it is commercially essential to get there, and the timescale for the change.

3   A detailed and comprehensive action plan for the development and implementation of the changes in working practices which explains in simple, easy to understand language what must be done differently by every member of the organisation. Adequate and appropriate training must support this for those that are required to operate in a different manner.

4   A simple and easy to understand explanation of the commercial benefits that will be delivered by the changes in working practices. This is best expressed in terms that relate to improved product quality, improved process efficiency, reduced waste in the production process and, most importantly, reduced costs and increased profits.

The Building Down Barriers case history in Chapter 4 provides considerable evidence that the adoption of the supply chain management principles that are common to the *Charter Handbook*, and to the *Modernising Construction* and *Better Public Buildings* Reports, will deliver major commercial improvements to both the demand and the supply side of the construction industry. Other major repeat clients in the UK, such as BAA, McDonald's and Argent, have actively taken control of their supply chains and used supply chain management techniques to deliver major commercial improvements for themselves and their supply-side partners. These include reductions in capital cost, better durability, fewer post-completion defects and less reworking. Not only does the adoption of supply chain management principles deliver improved lifetime quality and functionality and reduced capital and operating costs to the end user, the profits made by the supply-side firms will be enhanced by converting inefficiency and waste in the utilisation of labour and materials into lower costs and higher profits.

The 1994 Latham Report on the UK construction industry insisted that inefficiency and waste in the utilisation of labour and materials consumed at least 30 per cent of the initial capital cost of construction. The UK Building Services Research and Information Association (BSRIA) also confirmed this high level of unnecessary cost in 1997. Their Technical Note 14/97, *Improving Mechanical and Electrical Site Productivity*, provides well-documented evidence of the levels of efficiency in the utilisation of labour for mechanical and electrical services on projects in the UK, USA, Germany, France and Sweden. It also investigates the causes and advises on possible solutions, all of which involved better integration of the design and construction supply chain and greater involvement of specialist suppliers in design development. Recent work in the UK on a large number of projects for major repeat clients by the Building Research Establishment 'CALIBRE' team has also confirmed the Latham Report figure of 30 per cent for the level of unnecessary cost due to the inefficient utilisation of labour and materials.

It is highly unlikely that that inefficiency in the UK is restricted to new construction activities. Consequently, it is likely that construction activities on maintenance and refurbishment in the UK are also suffering a level of unnecessary cost due to the inefficient utilisation of labour and materials that amounts to 30 per cent of the initial capital cost.

Based on this overwhelming evidence, it follows that the adoption of the best practice of the three standards (*Charter Handbook*, *Modernising Construction* and *Better Public Buildings*) will eventually enable the construction industry to convert the unnecessary costs caused by inefficient utilisation of labour and materials into considerably lower prices for the industry's end user customers, and substantially higher profits for those firms in the industry's supply chain.

Case history no. 1: St Helens Metropolitan Borough Council

The Borough Council introduced a 'Best Value' approach to construction tendering as an alternative to the traditional confrontational approach of the industry with its 'sealed envelope' lowest price process which failed to provide clients with value for money because projects would regularly exceed the original cost and completion time predictions. Their 'Best Value' tendering approach appointed the entire design and construction supply chain from the outset, had a selection process that was based primarily on the skills and experience of the entire design and construction team (including the specialist suppliers), rather than the traditional 'lowest price' approach, and included sharing savings with the supply-side team on a 50–50 basis. When tested on a £2.3 million primary school, the out-turn cost to the Borough Council remained the same as the original cost estimate, but enhanced standards amounting to £400,000 were able to be built into the project. The school was completed ahead of the original planned completion date, much higher mechanical and electrical standards than the norm were achieved and the site had an excellent safety record. The Headteacher and his staff were involved in the partnering process from day one and were convinced that functionality was well above that usually delivered by new buildings.

Case history no. 2: Building Down Barriers pilot projects

The two pilot projects used as test-beds for the Building Down Barriers supply chain management toolset produced exceptional improvements in performance over conventionally procured buildings. Labour efficiency on site reached 70 per cent, which was almost double that usually achieved, and both specialist supplier teams came close to achieving a 'right first time' culture. Wastage of new materials was close to zero, whereas the industry norm is around 30 per cent. Both buildings were completed ahead of the contract completion date, with one being completed 20 per cent ahead. The through-life cost forecast for both buildings was well below that for traditionally procured buildings, without incurring a significantly increased capital cost. Functionality was demonstrated at every stage of the design process using 3-D visualisation, which avoided the usual problem of end users struggling to understand two-dimensional drawings, and the end users were delighted with actual functionality of the completed buildings which exceeded anything they had experienced before.

The USA *National Construction Goals* included a target for better design to deliver a 30 per cent improvement in occupant efficiency. Since it is unlikely that design in the UK is delivering significantly higher occupant efficiency levels than in the USA, a similar improvement in UK occupant efficiency ought to be possible if the integrated approach described in all three standards (*Charter Handbook*, *Modernising Construction* and *Better Public Buildings*) is adopted by the UK construction industry. The focus on delivering occupant efficiency is reflected strongly in the two key differentiators and the six primary goals of best practice listed at the beginning of this chapter. It does not require a great deal of logical thought to recognise the commercial benefit to the end user of buildings which enable the occupants to boost efficiency by 30 per cent, especially where the functional excellence and the environment is such that the morale of the end users is raised to a level where they are delighted to be working in or using the building. The *Better Public Buildings* Report makes the very telling point that, if hospitals provided a functionally excellent environment, patients would recover more quickly and would therefore be discharged earlier and this would represent a major operational saving for the hospital.

From the above evidence, it follows that there is an urgent and overwhelming need for a radical change in the sequential, fragmented procurement practice of clients and end users. All three standards (*Charter Handbook*, *Modernising Construction* and *Better Public Buildings*) see the consequences of sequential procurement as the cause of poor value from the construction industry. Ideally, this change should start with clients and end users willingly and enthusiastically embracing the two differentiators and the six goals of best practice procurement that are the foundation of the three standards. This also applies to those professional advisers that provide an interface between the end users and the construction industry, whether employed internally by the client within a property division or employed externally as consultants, since they normally provide the expert advice on procurement methods and, quite often, they place the design and construction contracts with the industry. It is the client's or end user's professional advisers that are usually responsible for the sequential appointment of the consultant designers and the construction contractors. Consequently, it is their actions and their advice to the client or end user that make it impossible for the specialist suppliers to have any input into the initial concept design, even though it is well known that most of the buildability problems on site are created in the first 20 per cent of the design process.

The publication of three well-defined and matching standards for best practice procurement, especially the two key differentiators and the six primary goals of best practice that are their foundation, enables those clients (and their professional advisers) who have to accord with the standards because of the pressure of external audit, or because they wish to become a Confederation of Construction Clients 'Chartered Client', to objectively review their current procurement practices and assess how closely they accord with the best practice of the three standards. Where any aspect of current practice is found to be at variance with best practice, action will need to be taken to improve the defective procurement practices.

One aspect of some clients' traditional procurement practice that directly inhibits supply chain integration is the practice of treating design as an in-house function that can safely be separated from construction. Where this occurs, it is almost impossible for the specialist suppliers to be involved in design from the outset and, thus, it is almost impossible to harness their experience and skill to eliminate inefficiency and waste in the utilisation of labour and materials in order to achieve a 'right first time' culture on site. As a consequence, where in-house design continues to be retained by the client, it is very unlikely that they will ever achieve best value in construction procurement.

The manner by which the best practice of the three standards is introduced by clients will be different depending on the nature of the client's construction programme. The majority of clients are small and occasional procurers of construction and are not therefore in a position to become a major driver of the change process. The clients that are repeat procurers of construction have considerable leverage and can use their changed procurement practice to force the pace and direction of the change process. Excellent examples of this in the UK are BAA, with their framework contract approach; Defence Estates, with prime contracting; and NHS Estates, with their Principle Supply Chain Manager approach. All three involve the introduction of supply chain integration and management by means that are appropriate to each client, and all three believe that involving specialist suppliers in design from the outset and thus harnessing their considerable skill and experience while the design solution is still at concept stage will deliver better value.

The following action plans briefly describe a possible change process that could be adopted by repeat clients and by small and occasional clients:

Best practice action plan – repeat clients

- Assess whether you want, or need, the commercial benefits of procurement best practice. For instance, significantly lower capital costs that come from the elimination of inefficiency and waste in labour and materials utilisation without the long-term risk of early, unplanned failure of cheap, poor quality components or materials in order to reduce the materials element of the capital cost. Accurately predicted, significantly lower and risk-free running costs that can be built into long-term business plans. Reduced operating costs come from the impact of excellent functionality on the efficiency and morale of the workforce. Greater cost and time certainty that comes from the actual out-turn cost never exceeding the forecast out-turn cost and the building always being completed on time. Never make assumptions about the effectiveness of your current procurement practices; always seek evidence to prove what has been achieved. Check if your construction contractors measure the efficient utilisation of labour and materials throughout their site activities. Speak to specialist suppliers about the frequency and causation of disruption to their work on site. Undertake end user surveys to test the effective functionality of completed buildings. Compare out-turn costs and actual completion dates with the initial budget and the forecast completion dates of a selection of completed buildings. Seek advice from your facility management staff on the maintainability, durability and ease of operation of completed buildings and facilities.
- Assess the current level of understanding of the three best practice standards, the two key differentiators and the six primary goals across your organisation. Never make assumptions about what you think people know; always seek evidence to demonstrate what individuals really understand. Ensure the assessment is carried out consistently across your organisation by the use of a simplified, easy to understand version of the three standards as the common evaluative criteria for

all the assessments. Use an independent expert, who can demonstrate a comprehensive understanding of the three standards to assist the assessment process. Ensure the assessment includes all levels of your organisation, since the knowledgeable and enthusiastic support of the Chief Executive and the Directors will be of paramount importance in any change process.

- Compare current practices and processes with the two key differentiators and the six primary goals of the three best practice standards, i.e. does the current procurement process ensure the involvement of specialist suppliers from the outset of the design process? This is critical to a 'right first time' culture on site. If current practice starts with the appointment of an architect to develop the design, who then appoints a construction contractor? Who in turn appoints specialist suppliers at a time when it is far too late for them to have any influence on the design (especially the elimination of inefficiency and waste in the utilisation of labour and materials)? If this is the case, your current procurement practice is not the best practice of the three standards. It is therefore unlikely that the six primary goals will be delivered and you are inevitably paying a heavy commercial penalty. Always ensure that this comparison is done objectively and never make assumptions or accept anyone's opinion as fact. It may be best to seek the assistance of an independent expert, who can demonstrate a comprehensive understanding of the three standards, to assist in the comparative analysis.
- Once the comparative analysis has objectively established any variance between current practices and the best practice of the three standards, the next step is to set an improvement target for the organisation and to ensure that everyone fully understands what the target means, why the target is important to the organisation, what benefits the improvements will deliver, how the improvements will be measured, and how the target will affect his or her individual role.
- Set Key Performance Indicators (KPIs) which enable you to measure accurately the organisation's rate of improvement.

These need to make sense when compared with the two key differentiators and the six primary goals of the three best practice standards. For instance, set KPIs which measure: end user satisfaction with functionality; both the production of, and the accuracy of, the cost of ownership predictions of the design and construction teams; the rate of improvement in the on-site utilisation of labour and materials or which compare the construction team's predicted utilisation levels with the actual utilisation levels; when specialist suppliers actually become involved in design development, especially whether their skill, knowledge and experience is really being used to the greatest advantage by the designers; and the speed of introduction of single point procurement.

- Ensure all leaders, especially the head of the organisation, become knowledgeable crusaders and champions for best practice. This requires them to have a deep and consistent understanding of the six primary goals and to ensure that every word they utter, every action they take, and everything they write reinforces the change process. It also requires a mechanism to assess how well everyone in the organisation understands what is entailed in the change process and how their work will be affected by it.

- Communicate the improvement targets and the intended changes in current practices and processes to everyone within your own organisation and in those organisations with which you interface in the construction industry. This is to ensure that everyone (including everyone with whom you interface) fully understands where you are going, why you are going there and how you intend to get there. Those with whom you interface will then be able to work out how it affects their own practices and processes. It is imperative that the language used is such that everyone can understand the message; there must be no ambiguities and there must be some way of checking the understanding of the recipients. The onus is always on the sender of the message to use the most appropriate language for the recipient and the KISS (Keep It Simple, Stupid) principle should always apply.

- Only invite expressions of interest from those firms that are able to provide well-documented evidence that they have practices and processes in place that ensure delivery of the six primary goals. In particular, demand the names of those firms (including the design and other consultants, and particularly including specialist suppliers) that are already bound together within an integrated team in long-term strategic supply chain partnerships.
- Evaluate the expressions of interest against the six primary goals so that you only select the integrated supply chain teams that can ensure best value by offering well-documented evidence of best practice.
- Focus in-house resources on defining the business need in terms that are suitable for measuring and evaluating the output performance of the built solution.

Best practice action plan – occasional clients (large and small)

- Assess whether you want or need the commercial benefits of procurement best practice. For instance, significantly lower capital costs that come from the elimination of inefficiency and waste in labour and materials utilisation without the long-term risk of early, unplanned failure of cheap, poor quality components or materials in order to reduce the materials element of the capital cost. Accurately predicted, significantly lower and risk-free running costs that can be built into long-term business plans. Reduced operating costs that come from the impact of excellent functionality on the efficiency and morale of the workforce. Greater cost and time certainty that comes from the actual out-turn cost never exceeding the forecast out-turn cost and the building always being completed on time. Never make assumptions about the effectiveness of current procurement practices; always demand evidence to prove what has been achieved on other projects. For instance, have the designers or the construction contractors measured the efficient utilisation of labour and materials throughout the site activities? Have they undertaken end user surveys to test effective functionality? Have they compared out-turn costs with the initial budgets, and can they tell you how often the two match? Have they compared the actual completion dates with the original target completion dates, and can they tell you how often the two match? Have they any evidence from facility management staff of the maintainability, durability and ease of operation of completed buildings and facilities?

- If procurement best practice is commercially beneficial, invite expressions of interest from those firms that are able to provide well-documented evidence that they have practices and processes in place that ensure delivery of the six primary goals. In particular, demand the names of those firms (including the design and other consultants, and particularly including specialist suppliers) that are already bound together within an integrated team in long-term strategic supply chain partnerships.

- Evaluate the expressions of interest against the six primary goals so that you only select the integrated supply chain teams that can ensure best value by offering well-documented evidence of best practice.

- If professional advice is deemed necessary, ensure those offering the advice can prove their deep and comprehensive understanding of the three best practice standards, especially the very comprehensive National Audit Office Report, Modernising Construction. (The 'Design Build Foundation' at Reading University and the 'Construction Best Practice Programme' at the Building Research Establishment can both assist in this area.)

- Ensure that you constrain your own organisation to defining the business need in detailed functional terms that are suitable for measuring and evaluating the output performance of the business activities housed within the built solution. Do not be tempted to define your requirement in terms of built solutions, because to do so will transfer risk to yourself if it does not perform as efficiently as it should.

Self-assessment questionnaire for best practice clients

The following questions may assist the self-assessment process against each of the six primary goals of construction best practice. When answering each question against an individual goal, use the available evidence to assess the degree by which the criteria is met, i.e. 0 per cent, 10 per cent, 20 per cent, 30 per cent of the time, etc. When all the questions have been answered, take the average percentage to give the overall percentage compliance with the specific goal. When the questions against each goal have been dealt with in this manner, it will show where the strengths and weaknesses lie and will enable the improvement process to be targeted at the weakest areas of performance.

Functionality

- How frequently are structured and facilitated value management workshops, that include end users, used to define and prioritise the detailed functional or business needs?
- How frequently are structured and facilitated value engineering workshops used by the design and construction team to support and validate all design decisions and the selection of all the components and materials?
- How many of those involved in the procurement process are aware of the three standards (*Charter Handbook*, *Modernising Construction* and *Better Public Buildings*) and how many of them can briefly summarise the key aspects of each standard?
- How many of those involved in the procurement process are able to give some of the examples cited in *Better Public Buildings* to illustrate the critical importance of delighting the end users?
- How many of those involved in the procurement process can explain how maximum functionality would benefit the competitiveness of a typical end user?
- How many of those involved in the procurement process can set out an action plan for ensuring maximum functionality will be delivered?

Cost of ownership

- How many of those involved in the procurement process are aware of the Confederation of Construction Clients' publication, *Whole Life Costing – A Client's Guide*, and can they briefly summarise the main points of the guide?
- How many of those involved in the procurement process are aware of the *Component Life Manuals* produced by the Building Performance Group (which is linked to the Housing Association Property Mutual insurance company) and published by Spon Press and Blackwell Publishing, and can they briefly summarise the purpose and value of the manuals?
- How many of those involved in the procurement process can briefly explain why predicting the cost of ownership might benefit an end user's business planning process?
- How many of those involved in the procurement process can briefly explain the difference between predicting the cost of ownership and estimating the cost of ownership?
- How many of those involved in the procurement process can briefly explain why the *Charter Handbook* is concerned about defects during the usage of a building and how these could affect the end user's competitiveness?

Inefficiency and waste

- How many of those involved in the procurement process have any experience of working with construction teams that have measured the effective utilisation of labour and materials?
- How many of those involved in the procurement process can briefly describe what the 1994 Latham Report had to say about the level of inefficiency and waste in the construction industry, and do they believe that level of inefficiency and waste still exists?
- How many of those involved in the procurement process are aware of what the National Audit Office Report, *Modernising Construction*, had to say about the degree to which the industry measures the effective utilisation of labour and materials?

- How many of those involved in the procurement process can briefly explain why it might be important to measure the effective utilisation of labour and materials?
- How many of those involved in the procurement process believe the current level of the effective utilisation of labour and materials to be well below what is acceptable?

Specialist suppliers

- How many of those involved in the procurement process can briefly describe the benefits gained by the design process with the involvement of specialist suppliers from the outset?
- How many of those involved in the procurement process can offer evidence of significant gains that have come directly from the early involvement of specialist suppliers in design?
- How many of those involved in the procurement process are aware of the Reading Construction Forum publication, *Unlocking Specialist Potential*, or the Building Services Research and Information Association Technical Note 14/97, *Improving M and E Site Productivity*?
- How many of those involved in the procurement process can suggest how or why the early involvement of specialist suppliers might deliver 'right first time' on site?
- How many of those involved in the procurement process can suggest and describe a form of procurement that would ensure the active involvement of key specialist suppliers from the outset of design development?

Single point of contact

- How many of those involved in the procurement process can give reasons why a single point of contact for the procurement of a building might aid the integration of design and construction?
- How many of those involved in the procurement process can briefly describe the likely barriers to procuring through a single point of contact?

- How many of those involved in the procurement process can explain how a single point of contact for the design and construction team would support effective supply chain management?
- How many of those involved in the procurement process have any direct experience of single point of contact procurement and can they briefly explain why single point of contact procurement is favoured by all three standards (*Charter Handbook*, *Modernising Construction* and *Better Public Buildings*)?
- How many of those involved in the procurement process believe that single point of contact is the only form of procurement that would ensure the involvement of specialist suppliers from the outset of design development?

Measurement

- How many of those involved in the procurement process can explain why all three standards see performance measurement as key to improvement?
- How many of those involved in the procurement process can explain why performance measurement is fundamental to an effective improvement programme?
- How many of those involved in the procurement process can give an example of an existing performance measurement system that has been tested, validated and approved by major repeat clients?

To be a best practice client, the traditional assumption that selection of the design and construction team based on lowest price gives best value needs to be radically reappraised, since the assumption has rarely proved true in reality. There is overwhelming evidence to show that the tender price is invariably exceeded when the final account is presented, and in many cases the final account has exceeded the tender price by over 60 per cent. This fact was emphasised in the National Audit Office Report, *Modernising Construction*, which stated:

In 1999, a benchmarking study of 66 central government departments' construction projects with a total value of £500 million
showed that three-quarters of the projects exceeded their budgets
by up to 50 per cent and two-thirds had exceeded their original
completion date by 63%.

The fact that the final settlement invariably exceeds the tender
price by a considerable margin when the contract is awarded on
the basis of the lowest price has been true for well over a hundred
years if John Ruskin is to be believed. In 1860, he said: 'If you deal
with the lowest bidder it is as well to add something for the risk you
run, and if you do that, you have enough to pay for something
better.' The UK Prime Minister, Margaret Thatcher, said in the
run-up to the 1979 election: 'If you pay peanuts, you get monkeys'.
Although she was talking about the civil service at the time, and
was emphasising the importance of paying salaries that would
attract high calibre people, the expression was taken up by the
wiser heads in the UK construction industry who have consistently
argued that lowest price tendering rarely delivers value for money
when the final settlement is rendered, and the picture becomes even
worse when whole-life costs are taken into account.

It therefore follows from the above that, for clients who are
determined to get better value from procurement, especially better
value in whole-life cost terms, selecting the design team separately
from the construction team and the use of the lowest price tendering must be replaced with some form of value-based selection that
appoints the entirety of the design and construction supply chain
from the outset on the basis of their skill and experience. This
moves the selection process from a simplistic focus on lowest price
to a much more sophisticated analysis of the skill and experience
of each member of the design and construction supply chain. It
also requires the introduction of a more open-book and trusting
relationship between the client and the integrated supply-side team
that is founded on performance measurement and a partnering
philosophy.

In the UK, government policy is that procurement should be
on the basis of value for money and not lowest cost. The UK

Minister for Trade and Industry has stated: 'The obsession with getting the lowest price for construction projects wastes money and cheats communities. Lowest cost does not mean better value.'

The more enlightened repeat clients in the UK have recognised that lowest price tendering has all too often turned out to be high risk to the commercial effectiveness of their business, because of the unpredictable price escalation that invariably occurs between the tender stage and the final settlement stage. They are exploring alternative ways of selecting, and awarding the contract to, an integrated supply-side team that places more reliance on trust, team working, open book and the sharing of savings and risks. BAA, Argent and McDonald's are good examples of this in the UK private sector, and Defence Estates and St Helens Metropolitan Borough Council are good examples in the UK public sector. The private sector clients that are moving away from tendering based on the lowest price tend to be major retailers who are using their own skill and experience of managing a retail supply chain to actively manage their design and construction supply chain through various forms of supply chain partnerships. Whilst this approach by major retailers is both valid and low risk for clients that possess a high level of supply chain management skill and experience from their retail or manufacturing side, it may be high risk and difficult for clients that lack this skill or cannot afford the skilled resources that would be necessary to actively manage the design and construction supply chain.

A reality of construction procurement is that in almost every case, the client and the end user have a budget for the construction works that represents the maximum affordable price for constructing a building or facility to house the relevant functional or business activities. This may come from the business planning process that is an essential part of assessing the affordability of the venture in a competitive market, or it may simply come from the amount of money available from savings, grants or a loan. In either case, if the final capital cost exceeds the maximum affordable price the client will suffer some form of financial hardship and, in the worse case (especially if the difference is over 50 per cent, as the National Audit Office found it was in three out of four public sector projects), this may well cause serious damage to the client's business competitiveness. This is equally true of the long-term operational and maintenance costs; the end user client cannot afford unpleasant and costly financial surprises at any stage in the operational life of the building or facility because they will inevitably have to be paid for out of their profit margin and will therefore adversely affect their commercial competitiveness.

To avoid the consequences of 'employing monkeys', which can involve serious budget overruns, poor whole-life quality in terms of durability and poor functionality, the client needs to select an integrated design and construction team that is able to prove they have the skill and experience to deliver a construction solution that efficiently satisfies the client's business or functional needs and does not exceed the client's maximum affordable price.

The starting point for any repeat client that wishes to ensure their procurement of construction is consistently delivering best value in capital cost, whole-life cost or functional terms, is to measure how well their current procurement process has performed. To do this they need to review the constructed outputs of their procurement process in terms of the final capital cost and the performance in use of the completed building or constructed facility. This requires an objective assessment of past construction works contracts to compare final settlements with tender prices. (How often did the final settlement match the tender price? By how much did the final settlement exceed the tender price in the worst case? What is the average escalation in cost between the tender price and the final settlement? How often have the construction team been required to measure the effective utilisation of labour and materials? What is the average level of the effective utilisation of labour and materials?) It requires an objective assessment of unexpectedly early component or materials failures during the planned life of the building or facility and an assessment of the impact on the commercial well-being of the end user of making good the failure. It also requires an objective assessment of the functional performance of the building or constructed facility by carefully interrogating the end users.

The situation is obviously different for small and occasional clients, since they do not have a continuous stream of construction contracts on which they can assess the effectiveness of their procurement process. Nevertheless, they would be well advised to ask their professional advisers about the true performance of the procurement approach they recommend. Similarly, they would be well advised to ask any potential construction contractor (including management contractors and design and build contractors) about their effectiveness in delivering value for money. Questions that would quickly paint a true picture are as follows:

- How often have final settlements matched tender prices?
- Where final settlements have exceeded tender prices, what was the worst case?
- What has been the average escalation in cost between tender price and final settlement?
- How often is the effective utilisation of labour and materials measured?
- What has been the average level of the effective utilisation of labour and materials in recent construction works?
- What evidence have they of the true durability (performance in use) of the components and materials used in completed buildings or facilities?
- How often have they provided an accurate prediction of the annual cost of ownership of a building or facility?
- What evidence have they of end user satisfaction with the functional performance of their completed buildings or facilities?

Any consideration of the appropriate procurement approach ought to bear firmly in mind what both the Latham and the Egan Reports made clear: that design must be fully integrated with construction if clients are to achieve better value from construction procurement. The Egan Report, in particular, emphasised that the route to better performance by the industry and thus to better value for clients, was for clients to buy their built products in the same way they bought their manufactured products. This very clearly requires design to be an intrinsic part of the supply side and to be totally integrated with construction using appropriate supply chain management techniques. Buying a built product in the same way as a manufactured product and use by the supply side of supply chain management techniques also requires the supply side to deal with the client through a single point of contact.

The following section suggests possible approaches to the selection and appointment of the fully integrated design and construction team.

Tendering by value to select an integrated design and construction team

A possible way forward in the selection and appointment of an integrated design and construction team that ensures the focus is fully on the skill and experience of the team is for the client to declare from the outset the maximum affordable price (ideally in both initial capital cost and long-term operational and maintenance cost terms). To ensure the delivery of best value, the client would need to stipulate that capital cost savings are to be shared on an equable basis (say 50–50); stipulate the use of an open-book approach throughout the entire design and construction supply chain (this needs to include transparency of profit margins and overheads); and stipulate that improvement in the effective utilisation of labour and materials must be a priority of the supply-side team and must be proven by measurement (this needs to include a requirement that the savings this exposes at specialist supplier level are included in the capital cost sharing arrangement). Having set the price to one side, the selection can be made on the basis of the skill and experience of each competing team using a carefully structured questionnaire. A suggested questionnaire based on the six goals of construction best practice is given in the next section (see pages 72–7).

Public sector clients in the UK are able to benefit from the work of pioneering best practice clients such as St Helens Metropolitan Borough Council, who have developed and successfully tested a two-part approach for the selection of an integrated design and construction team that gives a heavy weighting to the skill and experience aspect on the basis that it is this that will deliver the best value to the client and end user. As can be seen in the case history described earlier in this chapter (see page 53), the St Helens Metropolitan Borough Council approach delivered considerable benefits for all involved.

There are obviously other approaches to selection that are equally valid, but the client should ensure that the process used is for the selection of the fully integrated design and construction team advocated by the Egan Report and the three standards (*Charter Handbook*, *Modernising Construction* and *Better Public Buildings*). Ideally a two-envelope approach should be used so that the price can be dealt with separately from the skill and experience of the integrated team. Since the delivery of best value is most likely to come from the selection of the best design and construction team (in particular the selection of the best specialist suppliers), the marking system used should award the majority of the available marks to the skill and experience of the integrated team and should do so on the basis of the evidence of performance each team submits.

As suggested on page 69, the evidence submitted by the integrated design and construction supply-side team ought to include their recent achievements in delivering best value and at the very least ought to include the following:

- How often have final settlements matched tender prices?
- Where final settlements have exceeded tender prices, what was the worst case?
- What has been the average escalation in cost between tender price and final settlement?
- How often is the effective utilisation of labour and materials measured?
- What is the average level of the effective utilisation of labour and materials in recent construction works?
- What evidence have they of the true durability (performance in use) of the components and materials used in completed buildings or facilities?
- How often have they provided an accurate prediction of the annual cost of ownership of a building or facility?

- What evidence have they of end user satisfaction with the functional performance of their completed buildings or facilities?

The above questions, which relate to the delivery of best value for other clients, would provide a minimum assurance that the selected team have a good track record in the necessary skills needed to deliver best value on the current contract. However, where the achievement of best value is a commercial priority for the client, it might be wise to require the supply-side team to respond to a much more probing set of questions that could give greater assurance of their ability to deliver best value in terms of whole-life performance and the elimination of unnecessary costs. Basing the questionnaire on the six primary goals of construction best practice would ensure that all the necessary skills were covered and would bring a degree of uniformity to the selection process. Such a questionnaire is suggested below.

Questionnaire for assessing the skill and experience of an integrated design and construction team

In order to bring a degree of commonality across the industry to that part of the selection process that evaluates skill and experience of the entire supply-side design and construction team, it would make sense to include a series of questions that relate to the six primary goals of construction best practice from the Construction Best Practice Programme booklet, *A Guide to Best Practice in Construction Procurement*. It would also be essential to make the selection process as fair and objective as possible by basing the marking system on tangible evidence of experience, practice or performance rather than anecdotal claims. The following questionnaire may help the development of a value based selection system, although it may need to be adjusted depending on the nature or size of the individual construction works.

When dealing with the response to a question against an individual goal, use the available evidence to assess the degree by which the criteria is met by all those involved in the design and construction process, i.e. 0 per cent, 10 per cent, 20 per cent, 30 per cent of the time, etc. When the responses to the individual questions have been assessed, take the average of the percentage awarded to each question to give the overall percentage compliance with the individual goal. When each goal has been dealt with in this manner, the assessment will show where the strengths and weaknesses of the competing design and construction teams lie and will enable the selection decision to be fair and objective. The decision will also stand up to audit, whether the audit be internal or external.

A possible approach might be to issue the questionnaire to each competing design and construction team and ask them to submit a report that provides the response to each question, with examples of the evidence that supports the response. For instance, in the case of the first question on 'Functionality', they might state that the frequency was 20 per cent of all projects over the last five years and include one or two value engineering workshop reports from representative projects with attendance lists for the workshops that include end users, designers, construction contractors and specialist suppliers. In the case of the first question against 'Cost of ownership', they might state that the frequency was 5 per cent of all projects over the last five years and include one or two cost of ownership predictions from representative projects.

It would be essential to make clear to all the supply-side design and construction teams that are asked to make submissions that the term 'design and construction team' in the questionnaire refers to the entire supply chain and must therefore include the specialist suppliers.

Functionality

- How frequently are structured and facilitated value management workshops that include end users, designers, construction contractors and specialist suppliers used to define and prioritise the detailed functional or business needs?
- How frequently are structured and facilitated value engineering workshops used by the design and construction teams to support and validate all design decisions and the selection of all the components and materials?

- How many of those involved in the design and construction team have received training to assist their understanding of the three standards (*Charter Handbook*, *Modernising Construction* and *Better Public Buildings*)?
- How many of those involved in the design and construction team have received training to assist their understanding of the critical importance of delighting the end users?
- How many of those involved in the design and construction team have received training to assist their understanding of how maximum functionality would benefit the competitiveness of a typical end user?

Cost of ownership

- How often is the client provided with a cost of ownership prediction?
- How many of those involved in the design and construction team regularly use the Confederation of Construction Clients' publication, *Whole Life Costing – A Client's Guide*?
- How many of those involved in the design and construction team regularly use the *Component Life Manuals* produced by the Building Performance Group (which is linked to the Housing Association Property Mutual insurance company and published by Spon Press and Blackwell Publishing) to ensure maximum whole life durability?
- How many of those involved in the design and construction team have received training to assist their understanding of why predicting the cost of ownership might benefit an end user's business planning process?

Inefficiency and waste

- How many of those involved in the design and construction team have any experience of measuring the effective utilisation of labour and materials?
- How many of those involved in the design and construction team have received training to assist their understanding of

what the 1994 Latham Report had to say about the level of inefficiency and waste in the construction industry, and to assist their understanding of the level of inefficiency and waste that still exists?

- How many of those involved in the design and construction team have received training to assist their understanding of what the National Audit Office Report, *Modernising Construction*, had to say about the degree to which the industry measures the effective utilisation of labour and materials?
- How many of those involved in the design and construction team have received training to assist their understanding of why it might improve the competitiveness of all supply-side firms to measure the effective utilisation of labour and materials?

Specialist suppliers

- How often are specialist suppliers involved in design from the outset?
- How many of those involved in the design and construction team have received training to assist their understanding of the benefits that the involvement of specialist suppliers from the outset could bring to the design process?
- How many of those involved in the design and construction team have received training to assist their understanding of the significant gains that would come from the early involvement of specialist suppliers in design?
- How many of those involved in the design and construction team have received training to assist their understanding of the Reading Construction Forum publication, *Unlocking Specialist Potential*, or the Building Services Research and Information Association Technical Note 14/97, *Improving M and E Site Productivity*?
- How many of those involved in the design and construction team have received training to assist their understanding of why the early involvement of specialist suppliers might deliver 'right first time' on site?

Single point of contact

- How many of those involved in the design and construction team have received training to assist their understanding of why a single point of contact to represent the design and construction supply side might aid the integration of design and construction?
- How many of those involved in the design and construction team have received training to assist their understanding of the likely barriers to procuring through a single point of contact?
- How many of those involved in the design and construction team have received training to assist their understanding of why a single point of contact for the design and construction team would support effective supply chain management?
- How many of those involved in the design and construction team have received training to assist their understanding of why single point of contact procurement is favoured by all three standards (*Charter Handbook*, *Modernising Construction* and *Better Public Buildings*)?

Measurement

- How many of those involved in the design and construction team have received training to assist their understanding of why all three standards see performance measurement as key to improvement?
- How many of those involved in the design and construction team have received training to assist their understanding of why performance measurement is fundamental to an effective improvement programme?
- How often do the design and construction team regularly record the actual man-hours worked (including abortive time) and the actual materials used (including wastage) and compare these with the forecast figures?
- How often do the design and construction team conduct a detailed analysis of abortive time and materials wastage, assess the causes and work with other team members to seek ways of eliminating those causes?

- How often do the design and construction team set improvement targets that are aimed at reducing (and eventually eliminating) the gap between the forecast labour and materials figures and the actual labour and materials figures?
- How often do members of the design and construction team open their books to other team members and share their detailed data on current and past labour and materials costs, their profits and their overheads?

Accelerating Change Report

The 'Accelerating Change' report published in September 2002 by the Strategic Forum for Construction and chaired by Sir John Egan will inevitably force the pace and direction of radical damage by stipulating that:

> By the end of 2004, 20 per cent of construction projects by value should be undertaken by integrated teams and supply chains; and, 20 per cent of client activity by value should embrace the principles of the Clients' Charter. By the end of 2007 both these figures should rise to 50 per cent.

7   Integrating the design and construction team

Converting unnecessary costs into lower prices for clients and higher and more assured profits for supply-side firms, and providing clients with tangible evidence of radically improved and integrated working practices are the two powerful and sound commercial reasons why firms on the supply-side of the construction industry ought to adopt the totally integrated approach described in the three best practice standards, namely the Confederation of Construction Clients' *Charter Handbook*, the National Audit Office Report, *Modernising Construction*, and the Department of Culture, Media and Sport Report, *Better Public Buildings*.

In Chapter 4, the need for the creation of 'virtual firms' of consultants, construction contractors and specialist suppliers was explained. These long-term supply-side partnerships are essential to the introduction of the supply chain management tools and techniques demanded by the 1998 Egan Report, *Rethinking Construction*, and to the elimination of waste in the utilisation of labour and materials demanded by the 1994 Latham Report, *Constructing the Team*.

Within the supply-side 'virtual firm', total integration of design and construction and the adoption of the seven principles of supply chain management will enable the conversion of the high level of unnecessary costs (that come from the inefficient utilisation of labour and materials) into far higher and more assured profits for all of the supply-side firms. This conversion will also enable the firms in the 'virtual firm' to consistently tender lower prices to the end user customer of constructed products and, as a consequence, will enable them to maintain or increase their market share. In short, by agreeing to work together within long-term strategic supply-side partnerships that embody all aspects of supply chain management, supply-side firms will be far stronger and far more commercially healthy than they would be if they continued to operate in the isolated and fragmented way the industry has traditionally operated.

The 1994 Latham Report insisted that the level of unnecessary costs generated by the inefficient utilisation of labour and materials was around 30 per cent of the initial capital cost. This figure has been validated by the systematic measurement of actual practice by both the UK Building Services Research and Information Association (BSRIA) and the UK Building Research Establishment (BRE). In the case of BSRIA, the evidence was captured in their Technical Note 14/97, *Improving Mechanical and Electrical Site Productivity*, which provided well-documented evidence of the levels of efficiency in the utilisation of labour for mechanical and electrical services on projects in the UK, USA, Germany, France and Sweden. It also investigated the causes of the inefficiency and advised on possible solutions, all of which involved better integration of the design and construction supply chain and greater involvement of specialist suppliers in design development. Recent work in the UK on a large number of projects for major repeat clients by the Building Research Establishment 'CALIBRE' team has also confirmed the Latham Report figure of 30 per cent for the level of unnecessary costs due to the inefficient utilisation of labour and materials.

Case history no. 1: CRINE

The 1996 North Sea Oil Industry CRINE (Cost Reduction in the New Era) was necessitated because oil prices slumped from $23 a barrel down to $13 and the capital and operational costs of oilrigs were a major factor in the cost of oil production in the North Sea.

The solution agreed by all the oil companies operating there was for individual oil companies to work closely with their specialist suppliers within a fully integrated supply chain, building on the trust that can be generated by long-term supply chain partnerships. The oil companies were convinced that this approach was the one most likely to deliver the greatest reduction in capital and operating costs, whilst enabling the supply-side firms to improve their profit margins. It would enable them to harness the very considerable knowledge and experience of the specialist suppliers to drive out the unnecessary costs in the effective utilisation of labour and materials. It would also enable them to maximise the use of standard components, which would further reduce capital and operating costs.

In the case of a painting contractor, the Chief Executive of the company said that his price for painting an oilrig prior to the introduction of supply chain management techniques had been £4 million. His profits were varied, frequently entailed a loss and quite often came from claims. His buyers bought the cheapest paint they could find and he paid his operatives as little as he could get away with. After the introduction of supply chain management techniques, his price for painting an oilrig went down to £2.8 million (a reduction of 30 per cent), but he now made a very comfortable and assured profit. His buyers now worked very closely with paint manufacturers to buy high quality paint with excellent durability and applicability characteristics. He employed fewer operatives, but he paid them a far higher wage in return for an end to reworking

Case history no. 2: Building Down Barriers

The Building Down Barriers test-bed pilot projects demonstrated precisely the same commercial benefits to both the specialist supplier and the upstream customers of the specialist supplier's product. In the case of a steel fabrication firm, the input of the skill, knowledge and experience of the erectors and the fabricators into design development from the outset ensured that the steel frame was erected 'right first time' on site with no reworking and no wastage of materials. This produced a much lower price to the construction contractor, but it also produced much higher profits for the steel fabrication firm. It also boosted the morale of the erectors and the fabricators because (for the first time) they were being properly valued and given the opportunity to make a very positive, direct contribution to the design of the building. Virtually all the specialist suppliers in the two test-bed pilot projects had the same experience and were convinced of the massive and sustained commercial benefits that would come to all firms on the supply side if they adopted true supply chain integration based on the seven universal principles of supply chain management within long-term supply-side partnerships.

Throughout the entire UK construction industry, the annual expenditure on all construction activities is around £58 billion. It therefore follows from the Latham Report findings, and from subsequent work by BSRIA and BRE, that the amount of unnecessary cost generated by the inefficient utilisation of labour and materials is around £7 billion each year (30 per cent of the construction cost). It does not require rocket science to realise that the commercial benefits of converting this very high level of unnecessary cost into much higher profits and much lower prices are massive. Admitting to the existence of this high level of unnecessary cost is not easy for any supply-side firm, partly because of the deep-seated adversarial attitudes of the industry and partly because of the resistance to the use of performance measurement. As the National Audit Office Report, *Modernising Construction*, pointed out, the industry rarely measures the effective utilisation of labour and materials and, as a consequence, it is forced to rely on subjective opinion rather than objective fact. Unfortunately, when everyone within an industry has traditionally adopted the same subjective assessment of performance, it becomes the industry norm and is assumed to represent objective fact, even though it is nothing of the kind.

Any supply-side firm that denies the Latham Report figure of 30 per cent as representing the initial capital cost consumed by the inefficient utilisation of labour and materials, should ask themselves what hard evidence they have to support any claim they make about their own, and their supplier's efficiency. Have they measured and recorded the amount of new materials that go off site as waste because of reworking? Have they measured and recorded the man-hours of operative time wasted because drawings are wrong, materials are not delivered on time or the previous activity has not finished as programmed? Have they measured and recorded the man-hours of operative time consumed by having to return to site to correct defective workmanship or defective materials (in terms of both in-contract and post-contract defects)? Have they measured and recorded the man-hours of operative time consumed by modifications to the building necessitated by errors or omissions in the project briefing process?

Apart from the direct benefit of higher and more assured profits for supply-side firms, the other reason for change will increasingly come from the demand from public sector clients for evidence to prove that the supply-side has made real and sustained changes to integrate design and construction in accordance with the three best practice standards. Public sector clients in central government, health authorities and local government are responsible for 40 per cent of the total construction expenditure in the UK. They are all subject to regular external audit and the two bodies that conduct the audits (the National Audit Office for central government bodies and the Audit Commission for health and local government bodies) will increasingly use the National Audit Office Report, *Modernising Construction*, and the Department of Culture, Media and Sport Report, *Better Public Buildings*, as the common evaluative criteria for their audit assessments. As a consequence, those public sector clients that are subject to external audit will have little option but to adopt the approach to best practice that is being used as a benchmark by the audit teams. It therefore follows that public sector clients will be under great external pressure to adopt the best practice approach that is common to both *Modernising Construction* and *Better Public Buildings*.

At its simplest, this means the procurement practices of all public sector clients will increasingly have to embrace the two differentiators and the six goals of procurement best practice described in the Construction Best Practice Programme booklet, *A Guide to Best Practice in Construction Procurement*.

The two key differentiators of construction best practice are:

1   **Abandonment of lowest capital cost as the value comparator.** This is replaced in the selection process with whole-life cost and functional performance as the value for money comparators. This means that industry must predict and be measured by its ability to deliver maximum durability and functionality (which includes delighted end users).
2   **Involving specialist contractors and suppliers in design from the outset.** This means abandoning all forms of traditional procurement which delay the appointment of the specialist suppliers (sub-contractors, specialist contractors and manufacturers) until the design is well advanced (most of the buildability problems on site are created in the first 20 per cent of the design process). Traditional forms of sequential appointment are replaced with a requirement to appoint a totally integrated design and construction supply chain from the outset. This is only possible if the appointment of the integrated supply chain is through a single point of contact – precisely as it would be in the purchase of every other product from every other sector.

The six primary goals of construction best practice are:

1   The finished building will ensure maximum functionality.
2   The end users will benefit from the lowest cost of ownership.
3   Inefficiency and waste in the utilisation of labour and materials will be eliminated.
4   The specialist suppliers will be involved in design from the outset to achieve integration and buildability.
5   The design and the construction of the building will be achieved through a single point of contact for the most effective co-ordination and clarity of responsibility.

6   Current performance and improvement achievements will be established by measurement.

Previous chapters explained the critical role of the Chief Executive in those firms that are intent on embracing the best practice of the three UK publications (*Charter Handbook*, *Modernising Construction* and *Better Public Buildings*). It was emphasised that it was of paramount importance for the Chief Executive to adopt a powerful crusading role if the change were to succeed. The Chief Executives of supply-side firms must ensure the following four ingredients are in place:

1   A clearly explained and rational goal which all can understand, with which all can identify, and which can be related to specific and rational improvements in performance that can be measured and compared with current performance. The supply-side firm's goal must obviously accord with the goals of the demand-side clients to ensure that both sides are heading in the same direction.

2   Committed, determined, consistent and overt leadership by the Chief Executive which leaves no one in any doubt about where the change must take the organisation, why it is commercially essential to go there, and the timescale for the change. It must be obvious to everyone within the firm that the changes in working practices are necessary to produce the evidence of best practice demanded by the external auditors of clients, particularly the National Audit Office and the Audit Commission.

3   A detailed and comprehensive action plan for the changes in working practices that explains in simple, easy to understand language what must be done differently by every member of the organisation. This action plan must start with the Chief Executive and must end with those at the front line delivering services or products to the customers. Again, it must be obvious to everyone within the firm that the changes in working practices are necessary to produce the evidence of best practice demanded by the external auditors of clients, particularly the National Audit Office and the Audit Commission. Adequate

and appropriate training must support this change process for all those that are required to operate in a different manner.

4 A simple and easy to understand explanation of the commercial benefits that will be delivered by the changes in working practices. This is best expressed in terms which relate to improved product quality, improved process efficiency, reduced waste in the production process and, most importantly, reduced costs and increased profits. The explanation also needs to include the commercial benefits that will be delivered to the end users of the building or constructed facility.

The publication of the three well-defined and matching standards for best practice procurement (*Charter Handbook*, *Modernising Construction* and *Better Public Buildings*), especially their two key differentiators and their six primary goals of best practice, creates a level playing field of understanding across both the demand side and the supply side of the industry. It enables supply-side firms and their Chief Executives to fully understand the changes to procurement practice that their end user clients will be introducing in the coming years. These changes in procurement practice will most probably start with the major repeat public sector clients because of the external audit pressure of the National Audit Office and the Audit Commission, but will inevitably include the major repeat private sector clients as they strive to adopt the best practice of the Confederation of Construction Clients' *Charter Handbook*. Where supply-side firms accept the two powerful and sound commercial reasons why they ought to adopt the totally integrated approach that is the basis of all three best practice standards, they will need to review objectively their current working practices to assess how closely they accord with that set out in the three standards. Where any aspect of current practice is found to be at variance with the best practice of the three standards, action will need to be taken to improve the defective construction practices.

In an industry as fragmented and adversarial as the construction industry, where the adversarial relationships are deeply embedded in the culture of the industry and even extend into the further education establishments and the numerous institutes and federations, driving forward radical change will inevitably be an uphill struggle for the individual firms. The fragmentation and adversarial relationships also beset the top of the industry's hierarchy and consequently the struggle is made worse by the lack of a universally agreed and unambiguous goal for the culture and structure of the industry. As a consequence, the need for Chief Executives to adopt a powerful crusading role is far higher than it would be in the manufacturing or retail sectors. It is imperative that they set and communicate in unambiguous terms the goal for the structure and culture of their own firm and that of their suppliers. Ideally, they need to do this in co-ordination with other Chief Executives so that those at the sharp end are not confused and demoralised by the existence of a multitude of different and possibly conflicting goals that appear to show an industry as fragmented and adversarial as it has always been.

When developing a change strategy it is of critical importance to remember that clients will increasingly demand to see tangible evidence of improved performance and will increasingly move away from the simplistic lowest price tendering that is now seen by a growing number of informed clients and the public sector auditors as the reason for poor whole-life value. Key to providing the evidence is measurement of performance and especially the measurement of the effective utilisation of labour and materials throughout the supply chain.

Whilst each Chief Executive must decide the approach most appropriate for their individual firm, the following bullet-point list of actions briefly describes a possible change process that could be adopted by any firm of any size in each of the various sectors of the supply side of the construction industry, namely the construction contractors, consultant designers, project managers and quantity surveyors, and the specialist suppliers.

Best practice action plan – construction contractors

Benchmark current practice within your firm and set a target for improvement

- Compare current practices and processes with the two key differentiators and the six primary goals of the three best practice standards, i.e. if you have no long-term strategic supply chain partnerships with your downstream suppliers; if whole-life costing, value management and value engineering are not a well documented part of normal practice; if you do not measure the effective utilisation of labour and materials; if you have no way of capturing and sharing the efficiency gains with the rest of the supply chain and the client, then your current practice is not the best practice of the three standards and is unlikely to deliver the evidence required by those clients that have adopted the six primary goals. Never make assumptions about what you think people know; always seek evidence to demonstrate what individuals really understand. Ensure the assessment is carried out consistently across your organisation by the use of a simplified, easy to understand version of the three standards as the common evaluative criteria for all the assessments. Use an independent expert, who can demonstrate a comprehensive understanding of the three standards and will be seen as someone without an axe to grind or an empire to build, to assist the assessment process. Ensure the assessment includes all levels of your organisation, since the knowledge-able and enthusiastic support of the Chief Executive and the Directors will be of paramount importance in any change process.
- Consider what evidence is available to support the accuracy of the labour and materials elements of the prices tendered by your firm, i.e. is detailed process mapping used to work out the flow of work stream activities and the man-hours required for each activity? Is there a means of checking actual man-hours worked against those anticipated in the process mapping and for apportioning cause for any deviation? Again, never assume

what is being done or accept verbal assurance that things are done in a particular way. Always demand hard evidence of what has been done on a regular basis in the past, this should ideally span the previous five years if it is truly an ingrained part of normal practice.

- Once the comparative analysis has objectively established any variance between current practices and the best practice of the three standards, the next step is to set an improvement target for the organisation and to ensure everyone fully understands what the target means, why the target is important to the organisation, what benefits the improvements will deliver, how the improvements will be measured, and how the target will affect his or her individual role.

- Set Key Performance Indicators (KPIs) which will enable you to accurately measure the organisation's rate of improvement. These need to make sense when compared with the two key differentiators and the six primary goals of the three best practice standards. For instance, set KPIs which require the increasing use of independent user surveys to measure end user satisfaction with functionality; which require the increasing production of, and increasing accuracy of, cost of ownership predictions by the design and construction teams for the end user clients; which measure the rate of improvement in the on-site utilisation of labour and materials or which compare the construction team's predicted utilisation levels with the actual utilisation levels; which measure when specialist suppliers actually become involved in design development, especially whether their skill, knowledge and experience is really being used to the greatest advantage by the designers; which measure the rate that your existing suppliers are brought into strategic supply chain partnerships; and which measure the speed at which single point procurement is offered by your firm's marketing division as the preferred way of doing business.

- Ensure all leaders, especially the head of your firm, become crusaders and champions for best practice. This necessitates their having a deep and consistent understanding of the six primary goals.

Upstream relationships with demand-side customers

- Compare the current procurement practice of each of your demand-side customers with the three best practice standards and their two key differentiators and their six primary goals, i.e. if whole-life costing, value management and value engineering are not required to be a well-documented part of normal practice; if they do not require you to measure the effective utilisation of labour and materials; if they do not require you to capture and share efficiency gains with themselves and with the rest of the supply chain; if they do not wish to see evidence that you ring-fence your supplier's profits at a reasonable level, then your customer's practice is not the best practice of the three standards.

- In the case of those demand-side customers that are intent on embracing the best practice of the three standards, set out to each of them your commitment to the best practice described in the three standards and to the six primary goals. Be honest about where you are in terms of the effective utilisation of labour and materials and explain how you have measured your current performance. Explain your improvement target and your (KPIs) and provide them with any evidence that is available of improvements and innovations that relate to the six primary goals.

- In the case of one-off, or occasional end user customers, use as the basis of your marketing approach to them your commitment to the best practice described in the three standards and your determination to deliver the evidence of radical improvement required by the six primary goals. Explain your improvement target and your (KPIs) and provide them with any evidence that is available of improvements and innovations that relate to the six primary goals. Focus particularly on the ability of the best practice approach to deliver cost savings in the initial capital cost as well as in the running costs, and to deliver improved functionality that will boost the efficiency and the delight of the end users. Explain and emphasise how these benefits will feed into their long-term business plans and will therefore improve their competitiveness by reducing their production costs.

- Where a demand-side customer shows an interest in forming a long-term strategic supply chain partnership with you in accordance with the seven principles of supply chain management, discuss and agree the basis of a long-term partnership that enables both sides to make a reasonable and assured profit and to share in the cost savings of any improvements or innovations. This arrangement should include all your down-stream suppliers that need to be part of the integrated team for the range of work the demand-side customer is likely to order and should be based on an open-book approach.

- In the case of those demand-side customers that intend to continue with their fragmented and adversarial relationship with their suppliers, start to consider whether you need their business.

Downstream relationships with suppliers

- Assess the current level of understanding of the three best practice standards, their two key differentiators and six primary goals within the firms that form your current down-stream supply chain. Never make assumptions about what you think people know, always seek evidence to demonstrate what individuals really understand. Ensure the assessment is carried out consistently across your suppliers by the use of a simplified, easy to understand version of the three standards as the common evaluative criteria for all the assessments. Use an independent expert, who can demonstrate a comprehensive understanding of the three standards, to assist the assessment process. Ensure the assessment includes the attitudes of the Chief Executives and the senior managers, since the knowledgeable and enthusiastic support of the Chief Executive and the senior managers will be of paramount importance in any change process.

- Compare their current practices with the six primary goals, i.e. if they show no interest in forming long-term strategic supply chain relationships with you which embrace the seven principles of supply chain management; if whole-life costing, value management and value engineering are not a well-documented part of their normal practice; if they do not measure the effective utilisation of labour and materials; if they have no way of capturing and sharing efficiency gains with yourself or the rest of the supply chain, then their current practice is not the best practice of the three standards. They will therefore have difficulty in working with you to deliver the evidence required by those clients that have adopted the six primary goals and will lack the culture and practices necessary to form good strategic supply chain partnerships. Always ensure that this comparison is done objectively and never make assumptions or accept anyone's opinion as fact. It may be best to seek the assistance of an independent expert, who can demonstrate a comprehensive understanding of the three standards, to assist the comparative analysis.

- Consider what evidence is available to support the accuracy of the labour and materials elements of the prices tendered by each of your suppliers, i.e. is detailed process mapping used to work out the flow of work stream activities and the man-hours required for each activity? Is there a means of checking actual man-hours worked against those anticipated in the process mapping and apportioning cause for any deviation?
- Having assessed your current suppliers, decide on a short list of those that are most likely to support your drive for best practice. Choose from this short list those firms with which you wish to set up long-term strategic supply chain partnerships.
- If you have the assurance of your supplier's commitment to you that can only come from within long-term strategic supply chain partnerships, consider how large your down-stream supply chain needs to be, i.e. you will need to look back over your previous contracts and break down your typical workload into services, components and materials such that you can recognise patterns of commonality. Future workload trends are highly likely to follow historic patterns. As a

consequence, you can use this commonality to set the down-stream supply chain partnerships that would give greatest value. You may only need two firms for a given component or service, unless there are particular geographical reasons for a greater number. The important consideration is that you should restrict the number of suppliers to those your historic workload indicates you can provide a reasonable flow of work. Without this down-stream workload assurance, it will not be worthwhile them making the investment necessary to embrace the best practice of the three standards. You also need to bear in mind the resource implications to yourself that will come from the need to work closely with each of your suppliers to develop and monitor mutually agreed improvement targets.

- You will also need to consider whether others, not normally in your supply chain, could add greater value by being included, i.e. you will need look back and examine where errors in design have caused reworking and consider who needs to be included to eliminate such errors in the future.

- Discuss and agree improvement targets with each of your chosen supply chain partners that will enable them to deliver the six primary goals. Ensure everyone within these firms, and everyone within firms with which they have an interface, fully understands what the target means and how their individual role will be affected.

- Discuss and agree KPIs that they can use to measure accurately their firm's rate of improvement. These must dovetail with your KPIs to ensure consistency in the overall improvement regime. For instance, set KPIs which require the increasing use of independent user surveys to measure end user satisfaction with functionality; which require the increasing production of, and increasing accuracy of, cost of ownership predictions by the design and construction teams for the end user clients; which measure the rate of improvement in the on-site utilisation of labour and materials or which compare the construction team's predicted utilisation levels with the actual utilisation levels; which measure when specialist suppliers actually become involved in design development, especially whether their skill, knowledge and experience is really being used to the greatest advantage by the designers; which measure the rate that your existing suppliers are brought into strategic supply chain partnerships; and which measure the speed at which single point procurement becomes the firm's preferred way of doing business.

- Set up arrangements to regularly monitor each supply chain partner's performance against their KPIs. This should include an open-book approach to costing and a willingness to share innovations and improvements with other firms within your supply chain. It should also include sharing your own firm's performance with your suppliers and opening your own books to them in a totally trusting and open partnering relationship.

- Ensure all leaders, especially the head of each firm, become crusaders and champions for best practice. This necessitates them having a deep and consistent understanding of the six primary goals and a commitment to their delivery.

Note: See the end of this chapter (pages 107–11) for suggested questions relating to each of the six primary goals of construction best practice that could be adapted to assess how close the current procurement practice of you and your suppliers accords with the best practice of the three standards (*Charter Handbook*, *Modernising Construction* and *Better Public Buildings*).

Best practice action plan – consultant designers, project managers and quantity surveyors

Benchmark current practice within your firm and set a target for improvement

- Compare current practices and processes with the two key differentiators and the six primary goals of the three best practice standards, i.e. if you have no long-term strategic supply chain partnerships with construction contractors or specialist suppliers; if whole-life costing, value management and value engineering are not a well documented part of normal practice; if you do not discuss measuring the effective utilisation of labour and materials with construction contractors and specialist suppliers; if you do not discuss ways of capturing and sharing the efficiency gains with the rest of the supply chain and the client, then your current practice is not the best practice of the three standards and is unlikely to deliver the evidence required by those clients that have adopted the six primary goals. Never make assumptions about what you think people know, always seek evidence to demonstrate what individuals really understand. Ensure the assessment is carried out consistently across your organisation by the use of a simplified, easy to understand version of the three standards as the common evaluative criteria for all the assessments. Use an independent expert, who can demonstrate a comprehensive understanding of the three standards, to assist the assessment process. Ensure the assessment includes all levels of your organisation, since the knowledgeable and enthusiastic support of the Chief Executive and the Directors will be of paramount importance in any change process.

- Once the comparative analysis has objectively established any variance between current practices and the best practice of the three standards, the next step is to set an improvement target for the organisation and to ensure everyone fully understands what the target means, why the target is important to the organisation, what benefits the improvements will deliver, how the improvements will be measured, and how the target will affect his or her individual role.

- Set Key Performance Indicators (KPIs) which will enable you to measure accurately the organisation's rate of improvement. These need to make sense when compared with the two key differentiators and the six primary goals of the three best practice standards. For instance, set KPIs which require the increasing use of independent user surveys to measure end user satisfaction with functionality; which require the increasing production of, and increasing accuracy of, cost of ownership predictions by the design and construction teams for the end user clients; which ensure the measurement of the rate of improvement in the on-site utilisation of labour and materials or which ensure the comparison of the construction team's predicted utilisation levels with the actual utilisation levels; which measure when specialist suppliers actually become involved in design development, especially whether their skill, knowledge and experience is really being used to the greatest advantage by the designers; and which measure the speed at which single point procurement is offered by your firm's marketing division as the preferred way of doing business.

- Ensure all leaders, especially the head of your firm, become crusaders and champions for best practice. This necessitates their having a deep and consistent understanding of the six primary goals.

Upstream relationships with demand-side customers (such as end users of buildings and facilities)

- Compare the current procurement practice of each of your demand-side customers (such as end users and their professional advisers) with the three best practice standards, their two key differentiators and six primary goals, i.e. if whole-life costing, value management and value engineering are not required to be a well-documented part of normal practice; if they do not require you to ensure that the construction team measures the effective utilisation of labour and materials; if they do not require you to ensure that the construction team captures and shares efficiency gains with the end users and the rest of the supply chain; if they do not require you to ensure that the construction contractor ring-fences their supplier's profits at a reasonable level, then your demand-side customer's practice is not the best practice of the three standards. Ensure the comparison is carried out objectively by the use of a simplified, easy to understand version of the three standards as the common evaluative criteria for all the assessments.

- In the case of those demand-side customers that are intent on embracing the best practice of the three standards, set out to each of them your commitment to the best practice encompassed by the two key differentiators and the six primary goals. Be honest about where you are in terms of establishing long-term strategic supply chain partnerships with construction contractors and specialist suppliers that are based on the seven Building Down Barriers principles of supply chain management. Explain your improvement target and your KPIs and provide them with any evidence that is available on improvements and innovations that relate to the two key differentiators and six primary goals.
- In the case of those demand-side customers that intend to continue with their fragmented and adversarial relationship with their suppliers, you might want to consider whether you need their business.

Upstream relationships with supply-side customers (such as design and build contractors)

- Compare the current procurement practice of your supply-side customers (such as design and build contractors) with the three best practice standards, their two key differentiators and six primary goals, i.e. if they have no long-term strategic supply chain relationships with your firm, if whole-life costing, value management and value engineering are not required to be a well documented part of normal practice; if they do not measure the effective utilisation of labour and materials; if they do not capture and share efficiency gains with themselves and the rest of the supply chain; if they do not ring-fence their supplier's profits at a reasonable level, then your supply-side customer's practice is not the best practice of the three standards and is unlikely to deliver the evidence required by those clients that have adopted the six primary goals. Ensure the assessment is carried out consistently by the use of a simplified, easy to understand version of the three standards as the common evaluative criteria for all the assessments.

- In the case of those supply-side customers that are intent on embracing the best practice of the three standards, set out to each of them your commitment to the best practice encompassed by the two key differentiators and the six primary goals. Be honest about where you are in terms of your current performance. Explain your improvement target and your KPIs and provide them with any evidence that is available on improvements and innovations that relate to the two key differentiators and six primary goals.
- Where the supply-side customer shows an interest in forming a long-term strategic supply chain partnership with you in accordance with the seven universal principles of supply chain management, discuss and agree the basis of a long-term partnership that enables both sides to make a reasonable and assured profit and to share in the cost savings of any improvements or innovations.
- In the case of those supply-side customers that intend to continue in their fragmented and adversarial relationship with their suppliers, you might want to consider whether you need their business, e.g. in the case of the Building Down Barriers pilot projects, a steel fabrication firm came to the conclusion that if it could secure six long-term strategic partnerships with enlightened construction contractors that were actively embracing the seven universal principles of supply chain management, the assured and ring-fenced profits from these partnerships would avoid the need to tender for any other work where profits were far more at risk.

Down-stream relationships with supply-side firms (such as construction contractors, specialist suppliers and other consultants)

- Assess the current level of enthusiasm for the three best practice standards, their two key differentiators and six primary goals within the firms with which you regularly work. Ensure the assessment is carried out consistently by the use of a simplified, easy to understand version of the three standards as the common evaluative criteria for all the assessments. You might need to use an independent expert, who can demonstrate a comprehensive understanding of the three standards, to assist the assessment process. Ensure the assessment includes the attitude of the Chief Executives, since the knowledgeable and enthusiastic support of the Chief Executive is of paramount importance in any change process.

- Compare their current practices and processes with the two key differentiators and the six primary goals, i.e. if they show no interest in forming long-term strategic supply chain relationships with you which embrace the seven principles of supply chain management; if whole-life costing, value management and value engineering are not a well documented part of their normal practice; if they do not measure the effective utilisation of labour and materials; if they have no way of capturing and sharing efficiency gains with yourself or the rest of the supply chain, then their current practice is not the best practice of the three standards. They will therefore have difficulty in working with you to deliver the evidence required by those clients that have adopted the six primary goals and will lack the culture and practices necessary to form good strategic supply chain partnerships. Always ensure that this comparison is done objectively and never make assumptions or accept anyone's opinion as fact.
- Having assessed those supply-side firms with which you regularly work, decide which are most likely to support your drive for the best practice of the three standards. Choose from this short list those firms with which you might wish to set up long-term strategic supply chain partnerships.
- Discuss and agree improvement targets with each of your chosen supply-side partners that will enable you to work together to deliver the evidence required by those clients that have adopted the six primary goals. Ensure everyone within your own firm, and everyone within those supply-side firms with which they have an interface, fully understands what the target means and how their individual role will be affected.

- Discuss and agree with your supply-side partners KPIs that you can both use to measure accurately the rate of improvement. These must all dovetail together to ensure consistency in the overall improvement regime. For instance, agree KPIs which require the increasing use of independent user surveys to measure end user satisfaction with functionality; which require the increasing production of, and increasing accuracy of, cost of ownership predictions by the design and construction teams for the end user clients; which measure the rate of improvement in the on-site utilisation of labour and materials or which compare the construction team's predicted utilisation levels with the actual utilisation levels; which measure when specialist suppliers actually become involved in design development, especially whether their skill, knowledge and experience is really being used to the greatest advantage by the designers; which measure the rate that existing suppliers are brought into strategic supply chain partnerships; and which measure the speed at which single point procurement becomes the firm's preferred way of doing business.

- Set up arrangements to regularly compare each firm's performance against the KPIs. This should include an open-book approach to costing and a willingness to share innovations and improvements with supply-side partners. It should also include being open about your own firm's performance and opening your own books to them in a totally trusting and open partnering relationship.

- Ensure all leaders, especially the head of each firm, become crusaders and champions for best practice. This necessitates them having a deep and consistent understanding of the six primary goals and a commitment to their delivery.

Note: See the end of this chapter (pages 107–11) for suggested questions relating to each of the six primary goals of construction best practice that could be adapted to assess how close the current procurement practice of your and other supply-side firms accords with the best practice of the three standards (*Charter Handbook*, *Modernising Construction* and *Better Public Buildings*).

Best practice action plan – specialist suppliers and manufacturers

Benchmark current practice within your firm and set a target for improvement

- Compare current practices and processes with the two key differentiators and the six primary goals of the three best practice standards, i.e. if you have no long-term strategic supply chain partnerships with your up-stream supply-side customers or down-stream suppliers; if whole-life costing, value management and value engineering are not a well-documented part of normal practice; if you do not measure the effective utilisation of labour and materials; if you have no way of capturing and sharing the efficiency gains with the rest of the supply chain and the demand-side end users, then your current practice is not the best practice of the three standards and is unlikely to deliver the evidence required by those clients that have adopted the six primary goals. Never make assumptions about what you think people know, always seek evidence to demonstrate what individuals really understand. Ensure the assessment is carried out consistently across your organisation by the use of a simplified, easy to understand version of the three standards as the common evaluative criteria for all the assessments. Use an independent expert, who can demonstrate a comprehensive understanding of the three standards, to assist the assessment process. Ensure the assessment includes all levels of your organisation, since the knowledgeable and enthusiastic support of the Chief Executive and the Directors will be of paramount importance in any change process.
- Consider what evidence is available to support the accuracy of the labour and materials elements of the prices tendered by your own firm and by your suppliers, i.e. is detailed process mapping used to work out the flow of work stream activities and the man-hours required for each activity; is there a means of checking actual man-hours worked against those anticipated in the process mapping and for apportioning cause for any

deviation. Again, never assume what is being done or accept verbal assurance that things are done in a particular way. Always demand hard evidence of what has been done on a regular basis in the past; this should ideally span the previous five years if it is truly an ingrained part of normal practice.

- Once the comparative analysis has objectively established any variance between current practices and the best practice of the three standards, the next step is to set an improvement target for the organisation and to ensure everyone fully understands what the target means, why the target is important to the organisation, what benefits the improvements will deliver, how the improvements will be measured, and how the target will affect his or her individual role.

- Set Key Performance Indicators (KPIs) which will enable you to measure accurately the organisation's rate of improvement. These need to make sense when compared with the two key differentiators and the six primary goals of the three best practice standards. For instance, set KPIs which require the increasing use of independent user surveys to measure end user satisfaction with functionality; which require the increasing production of, and increasing accuracy of, cost of ownership predictions by the design and construction teams for the end user clients; which measure the rate of improvement in the on-site utilisation of labour and materials or which compare the construction team's predicted utilisation levels with the actual utilisation levels; which measure when specialist suppliers actually become involved in design development, especially whether their skill, knowledge and experience is really being used to the greatest advantage by the designers; which measure the rate that your existing suppliers are brought into strategic supply chain partnerships; and which measure the speed at which single point procurement is offered by your firm's marketing division as the preferred way of doing business.

- Ensure all leaders, especially the head of your firm, become crusaders and champions for best practice. This necessitates their having a deep and consistent understanding of the six primary goals.

Up-stream relationships with supply-side customers
(such as construction contractors)

- Compare the current procurement practice of each of your supply-side customers (such as construction contractors) with the three best practice standards, their two key differentiators and six primary goals, i.e. if they have no long-term strategic supply chain relationships with your firm; if whole-life costing, value management and value engineering are not required to be a well-documented part of normal practice; if they do not require you to measure the effective utilisation of labour and materials; if they do not require you to capture and share efficiency gains with themselves and the rest of the supply chain; if they do not ring-fence their supplier's profits at a reasonable level, then your supply-side customer's practice is not the best practice of the three standards and is unlikely to deliver the evidence required by those clients that have adopted the six primary goals. Ensure the assessment is carried out consistently across your supplier's organisations by the use of a simplified, easy to understand version of the three standards as the common evaluative criteria for all the assessments.

- In the case of those supply-side customers that are intent on embracing the best practice of the three standards, set out to each of them your commitment to the best practice encompassed by the two key differentiators and the six primary goals. Be honest about where you are in terms of the effective utilisation of labour and materials and explain how you have measured your own and your supplier's current performance. Explain your improvement target and your KPIs and provide them with any evidence that is available of improvements and innovations that relate to the two key differentiators and the six primary goals.

- Where the supply-side customer shows an interest in forming a long-term strategic supply chain partnership with you in accordance with the seven principles of supply chain management, discuss and agree the basis of a long-term partnership that enables both sides to make a reasonable and assured profit and to share in the cost savings of any improvements or innovations.

- In the case of those supply-side customers that intend to continue in their fragmented and adversarial relationship with their suppliers, you might want to consider whether you need their business, e.g. in the case of the Building Down Barriers pilot projects, a steel fabrication firm came to the conclusion that if it could secure six long-term strategic partnerships with enlightened construction contractors that were actively embracing the seven universal principles of supply chain management, the assured and ring-fenced profits from these partnerships would avoid the need to tender for any other work where profits were far more at risk.

Up-stream relationship with demand-side customers (such as end users of buildings or facilities)

- In the case of those repeat end user customers with which you have a direct relationship, act as described above for supply-side customers.
- In the case of one-off, or occasional end user customers, use your commitment to the best practice described in the three standards and your determination to deliver the evidence of radical improvement required by the six primary goals as the basis of your marketing approach to them. Explain your improvement target and your Key Performance Indicators (KPIs) and provide them with any evidence that is available of improvements and innovations that relate to the six primary goals. Focus particularly on the ability of the best practice approach to deliver cost savings in the initial capital cost as well as in the running costs, and to deliver improved functionality that will boost the efficiency and the delight of the end users. Explain and emphasise how these benefits will feed into their long-term business plans and will therefore improve their competitiveness by reducing their production costs.
- Where the one-off, or occasional, end user customer shows no interest in the six primary goals or the three best practice

standards, you might want to decide if you need their business, since such work will inevitably have a high risk of poor profitability.

Down-stream relationships with suppliers

- Assess the current level of understanding of the three best practice standards, their two key differentiators and six primary goals in the firms that form your current down-stream supply chain. Ensure the assessment is carried out consistently by the use of a simplified, easy to understand version of the three standards as the common evaluative criteria for all the assessments. Use an independent expert, who can demonstrate a comprehensive understanding of the three standards, to assist the assessment process. Ensure the assessment includes all levels of your supplier's organisations, since the knowledgeable and enthusiastic support of the Chief Executive and the Directors will be of paramount importance in any change process.

- Compare their current practices and processes with the two key differentiators and the six primary goals, i.e. if they show no interest in forming long-term strategic supply chain relationships with you which embrace the seven principles of supply chain management; if whole-life costing, value management and value engineering are not a well documented part of their normal practice; if they do not measure the effective utilisation of labour and materials; if they have no way of capturing and sharing efficiency gains with yourself or the rest of the supply chain, then their current practice is not the best practice of the three standards. They will therefore have difficulty in working with you to deliver the evidence required by those clients that have adopted the six primary goals and will lack the culture and practices necessary to form good strategic supply chain partnerships. Always ensure that this comparison is done objectively and never make assumptions or accept anyone's opinion as fact.

- Consider what evidence is available to support the accuracy of the labour and materials elements of the prices tendered by each of your suppliers, i.e. is detailed process mapping used to work out the flow of work stream activities and the man-hours required for each activity? Is there a means of checking actual man-hours worked against those anticipated in the process mapping and apportioning cause for any deviation?

- Having assessed your current suppliers, decide which are most likely to support your drive for best practice. Choose from this short list those firms with which you might wish to set up long-term strategic supply chain partnerships.
- If you have the assurance of only working with suppliers that are as enthusiastically committed as your own firm to delivering the evidence required by the two key differentiators and the six primary goals of best practice and firmly believe that this evidence can only come from supply-side firms that are working within long-term strategic supply chain partnerships, consider how large your down-stream supply chain needs to be, i.e. you will need to look back over your previous contracts and break down your typical workload into services, components and materials such that you can recognise patterns of commonality. Future workload trends are highly likely to follow historic patterns. As a consequence, you can use this commonality to set the down-stream supply chain partnerships that would give greatest value. You may only need two firms for a given component or service, unless there are particular geographical reasons for a greater number. The important consideration is that you should restrict the number of suppliers to those your historic workload indicates you can provide a reasonable flow of work. Without this down-stream workload assurance, it will not be worthwhile their making the investment necessary to embrace the best practice of the three standards. You also need to bear in mind the resource implications to yourself that will come from the need to work with each of your suppliers in developing and monitoring mutually agreed improvement targets.
- You will also need to consider whether others, not normally in your supply chain, could add greater value by being included, i.e. you will need look back and examine where

errors in design have caused reworking and consider who
needs to be included to eliminate such errors in the future.

- Discuss and agree improvement targets with each of your
chosen suppliers that will enable them to deliver the evidence
required by those clients that have adopted the six primary
goals. Ensure everyone within their firm, and everyone within
those firms with which they have an interface, fully understands what the target means and how their individual role
will be affected.

- Discuss and agree KPIs that they can use to accurately measure
their firm's rate of improvement. These must dovetail with
your KPIs to ensure consistency in the overall improvement
regime. For instance, set KPIs which require the increasing use
of independent user surveys to measure end user satisfaction
with functionality; which require the increasing production
of, and increasing accuracy of, cost of ownership predictions by the design and construction teams for the end user
clients; which measure the rate of improvement in the on-site
utilisation of labour and materials or which compare the
construction team's predicted utilisation levels with the actual
utilisation levels; which measure when specialist suppliers
actually become involved in design development, especially
whether their skill, knowledge and experience is really being
used to the greatest advantage by the designers; which measure
the rate that existing suppliers are brought into strategic supply
chain partnerships; and which measure the speed at which
single point procurement becomes the firm's preferred way of
doing business.

- Set up arrangements to regularly monitor each firm's performance against their KPIs. This should include an open-book
approach to costing and a willingness to share innovations and
improvements with other firms within your supply chain. It
should also include sharing your own firm's performance with
your suppliers and opening your own books to them in a
totally trusting and open partnering relationship.

- Ensure all leaders, especially the head of each firm, become
crusaders and champions for best practice. This necessitates

them having a deep and consistent understanding of the six primary goals and a commitment to their delivery.

Suggested assessment questionnaire

The following questions may assist the assessment process relating to each of the six primary goals of construction best practice. When answering each question against an individual goal, use the available evidence to assess the degree by which the criteria is met by all those involved in the design and construction process, i.e. 0 per cent, 10 per cent, 20 per cent, 30 per cent of the time, etc. When all the questions have been answered, take the average percentage to give the overall percentage compliance with the specific goal. When the questions against each goal have been dealt with in this manner, the self-assessment will show where the strengths and weaknesses lie and will enable the improvement process within the firm to be targeted at the weakest areas of performance. The third party assessment will enable the selection process for supply chain partners to be based on objective, fairly compiled and well-structured evidence.

Functionality

- How frequently are structured and facilitated value management workshops, that include end users, designers, construction contractors and specialist suppliers, used to define and prioritise the detailed functional or business needs?
- How frequently are structured and facilitated value engineering workshops used by the design and construction teams to support and validate all design decisions and the selection of all the components and materials?
- How many of those involved in the design and construction teams are aware of the three standards (*Charter Handbook*, *Modernising Construction* and *Better Public Buildings*), and how many of them can briefly summarise the key aspects of each standard?

- How many of those involved in the design and construction teams are able to give some of the examples cited in *Better Public Buildings* to illustrate the critical importance of delighting the end users?
- How many of those involved in the design and construction teams can explain how maximum functionality would benefit the competitiveness of a typical end user?
- How many of those involved in the design and construction teams can set out and explain an action plan for ensuring maximum functionality will be delivered?

Cost of ownership

- How many of those involved in the design and construction teams are aware of the Confederation of Construction Clients' publication, *Whole Life Costing – A Client's Guide*, and can they briefly summarise the main points of the *Guide*?
- How many of those involved in the design and construction teams are aware of the *Component Life Manuals* produced by the Building Performance Group (which is linked to the Housing Association Property Mutual insurance company) and published by Spon Press and Blackwell Publishing, and can they briefly summarise the purpose and value of the *Manuals*?
- How many of those involved in the design and construction teams can briefly explain why predicting the cost of ownership might benefit an end user's business planning process?
- How many of those involved in the design and construction teams can briefly explain the difference between predicting the cost of ownership and estimating the cost of ownership?
- How many of those involved in the design and construction teams can briefly explain why the *Charter Handbook* is concerned about defects during the usage of a building and how these could affect the end user's competitiveness?

Inefficiency and waste

- How many of those involved in the design and construction teams have any experience of measuring the effective utilisation of labour and materials?
- How many of those involved in the design and construction teams can briefly describe what the 1994 Latham Report had to say about the level of inefficiency and waste in the construction industry, and do they believe that that level of inefficiency and waste still exists?
- How many of those involved in the design and construction teams are aware of what the National Audit Office Report, *Modernising Construction*, had to say about the degree to which the industry measures the effective utilisation of labour and materials?
- How many of those involved in the design and construction teams can briefly explain why it might be important to improved competitiveness to measure the effective utilisation of labour and materials?
- How many of those involved in the design and construction teams believe the current level of the effective utilisation of labour and materials to be well below what is acceptable?

Specialist suppliers

- How many of those involved in the design and construction teams can briefly describe the benefits the involvement of specialist suppliers from the outset could bring to the design process?
- How many of those involved in the design and construction teams can offer evidence of significant gains that have come directly from the early involvement of specialist suppliers in design?
- How many of those involved in the design and construction teams are aware of the Reading Construction Forum publication, *Unlocking Specialist Potential*, or the Building Services Research and Information Association Technical Note 14/97, *Improving M and E Site Productivity*?

- How many of those involved in the design and construction teams can suggest how or why the early involvement of specialist suppliers might deliver 'right first time' on site?
- How many of those involved in the design and construction teams can suggest and describe a form of procurement that would ensure the active involvement of key specialist suppliers from the outset of design development?

Single point of contact

- How many of those involved in the design and construction teams can give reasons why a single point of contact for the procurement of a building might aid the integration of design and construction?
- How many of those involved in the design and construction teams can briefly describe the likely barriers to procuring through a single point of contact?
- How many of those involved in the design and construction teams can explain how a single point of contact for the design and construction team would support effective supply chain management?
- How many of those involved in the design and construction teams have any direct experience of single point of contact procurement and can they briefly explain why single point of contact procurement is favoured by all three standards (*Charter Handbook*, *Modernising Construction* and *Better Public Buildings*)?
- How many of those involved in the design and construction teams believe that single point of contact is the only form of procurement that would ensure the involvement of specialist suppliers from the outset of design development?

Measurement

- How many of those involved in the design and construction teams can explain why all three standards see performance measurement as key to improvement?

- How many of those involved in the design and construction teams can explain why performance measurement is fundamental to an effective improvement programme?
- How many of those involved in the design and construction teams can give an example of an existing performance measurement system that has been tested, validated and approved by major repeat clients?
- How many of those involved in the design and construction teams regularly record the actual man-hours worked (including abortive time) and the actual materials used (including wastage) and compare these with the forecast figures?
- How many of those involved in the design and construction teams conduct a detailed analysis of abortive time and materials wastage, assess the causes and work with other team members to seek ways of eliminating those causes?
- How many of those involved in the design and construction teams set improvement targets that are aimed at reducing (and eventually eliminating) the gap between the forecast labour and materials figures and the actual labour and materials figures?
- How many of those involved in the design and construction teams are prepared to open their books to other team members and share their detailed data on current and past labour and materials costs, their profits and their overheads?

8 Ensuring the delight of the end users

A primary thrust of the UK Department of Culture, Media and Sport Report, *Better Public Buildings*, the National Audit Office Report, *Modernising Construction*, and the UK Confederation of Construction Clients' *Charter Handbook* is the need to ensure the functional performance of the constructed facility is such that the morale and efficiency of those working in and using the facility are enhanced. This primary thrust is also common to the USA *National Construction Goals* that strongly emphasises the need for the buildings and facilities constructed in the USA to enhance the competitiveness of their occupants and users. In fact, one of the seven USA *National Construction Goals* demands that the design of buildings be improved sufficiently to deliver a 50 per cent enhancement to the performance of the building occupants.

It is obvious that the functional performance and the morale of the end users can only be improved if the design is based on a thorough and detailed understanding of the precise functional requirements and of the many inter-related and prioritised values of the end users. The whole thrust of the demand for total integration of the design and construction team means that this understanding must embrace everyone on the supply-side whose knowledge and experience could be beneficial to the development of the design; this must obviously include the specialist suppliers and manufacturers. It is no longer sufficient for this understanding of the end user's functional requirements to be constrained to the architects and engineers that have traditionally produced the design, with the skilled craftsmen who construct the building on site having little or no idea of the detailed business needs of the end user that the completed building must satisfy.

As in other sectors, this understanding of the precise end user's functional requirements must embrace every member of the design and construction supply chain. The bricklayer, the carpenter and the electrician must have a clear and detailed understanding of the end user's functional requirements that must be satisfied if the completed building is to be deemed a success. It is imperative that those constructing the building associate their success with the delight of the end users that comes from the ability of the building to enable them to do their jobs smoothly and efficiently in an environment that boosts their morale.

A clear and detailed understanding of the end user's functional requirements and values by the specialist suppliers' and the manufacturers' operatives cannot be achieved if it is filtered through a long chain of communication. All too often, the end user requirements are initially filtered through construction professionals or procurement staff within the client organisation who interpret them and incorporate them into the project brief. They are then filtered again through the architect and the engineers, and then again through the construction contractor before they reach the specialist suppliers. The manufacturers then have the end user requirements filtered, yet again, by the specialist suppliers. In all, there may be five or more links in the supply chain filtering those original end user requirements before the manufacturers come into the design process.

The sad reality is that it is rare if any operative on site has any real understanding of the end user's detailed functional requirements and business values. The operatives may know what they are constructing in terms of its physical appearance and its form, but they are very unlikely to have any detailed knowledge or understanding of the functional activities that must be delivered efficiently by the finished building. Worse still, they will never receive feedback on how well the needs and aspirations of the end users were met by the completed building. In fact, it is rare if any member of the design and construction supply team receives objective feedback on the degree to which the completed building satisfies the functional requirements of the end users.

Clearly, the first thing that has to be done to overcome this deficiency is to integrate the design and construction supply chain in such a way as to ensure the specialist suppliers and the manufacturers that will be constructing the building or facility are available at the outset, when the end user's requirements are analysed and defined. The means by which the totality of the integrated supply chain can be on hand from the outset is dealt with elsewhere in this guide under the explanation of the formation and operation of the 'virtual firm' (see Chapter 4).

However, the mere existence of the 'virtual firm' is not sufficient to automatically ensure everyone in the design and construction team has a detailed understanding of the end user's requirements. This understanding can only come from the use of a formal, structured process to tease out, prioritise and define every end user's requirement that must be satisfied efficiently by the completed building. The resulting project brief will clearly be of a far higher and more consistent quality, and far less prone to error if the structured process has a well-established track record of success. The use of a formal, structured process is also essential if it is to be applied consistently across the industry by different teams for different end users. Operating 'by God and by guesswork' is not an acceptable way of working if there is to be any assurance that end users will always be delighted by the functional performance of the constructed building or facility.

A process tool for teasing out the totality of the detailed end user's functional requirements has been available for some considerable time in the form of Value Management, which is a structured approach to defining what 'value' means to an end user. It ensures that:

- The need for a construction project is always verified and supported by data.
- Project objectives are openly discussed, clearly identified and well defined.
- Key decisions are rational, explicit and accountable, and are focused on value rather than cost.

- The design evolves within an agreed framework of project objectives and seeks to achieve the best balance between time, cost and quality.
- All involved have a shared understanding.
- Unnecessary cost is eliminated.

All too often, those involved in traditional design and construction are guilty of using terms like 'Value Management' with very little evidence that a structured approach to value management has been used or of the degree to which end users or specialist suppliers have been involved and their values recorded. If the primary thrust of delighting the end user according to *Better Public Buildings*, *Modernising Construction*, *Charter Handbook* and the USA *National Construction Goals* is to be achieved, the traditional informal and unstructured approach to the production of the project brief by the client's professional advisers, in isolation from the specialist suppliers, must be replaced by the use of structured value management that includes end users and specialist suppliers. A skilled facilitator who has been trained in value management and can provide evidence of successful experience in facilitating value management workshops must lead this process. Ideally, this evidence should include feedback from end users on the effective performance of a building or facility which has been designed and constructed from the output of a formal, structured value management process which has been facilitated by the individual providing the evidence.

Case history: Building Down Barriers

The Building Down Barriers test-bed pilot projects used value management workshops to define the end user's requirements and values in considerable detail and with great precision. These workshops were organised and chaired by an experienced value management facilitator with a proven track record of success whose first task was to identify key end users and key supply chain members (especially those from the specialist suppliers). The end user representatives included key individuals from the policy division that were responsible for the long-term strategy and could advise on the likely future changes in use over the functional lifetime of the building. The outcome of the workshops was threefold in nature, namely:

- A very detailed and precise project brief that listed, prioritised and described every essential and desirable functional requirement. (In the case of one of a swimming pools, it set the Olympic standard as essential for the swimming and diving requirements.) A recognition by the supply-side that it is the efficiency of the functional activities housed by the building that are of paramount importance to the end users; the building itself was recognised to be of secondary importance and was seen as merely a weather-tight envelope that provided a suitable environment for the functional activities.

- A desire that those attending the value management workshops should also attend the value engineering workshops so that the selection of the design and component options was a team decision and the end user members of the team could understand why the decision was made. This was seen as important to the team because the value management workshops had taught the supply-side the importance of ensuring all design and component selections were made with the need to keep the building operational throughout its working life. The value management workshops had also taught them the financial consequences to the end users if any part of the building was out of use for any reason.

- The traditional split between the demand-side and supply-side disappeared during the value management workshops and was replaced by a single, enthusiastic team dedicated to the design and construction of a building that would efficiently deliver the full range of the end user's functional requirements and values. The team spirit that developed was mutually supportive, open and trusting. One high level visitor to the site of one of the two test-bed pilot projects said it was the only site he had ever visited where everyone he spoke to was enthusiastic, happy and totally convinced that the building they were constructing would perform exceptionally well in functional terms.

This structured process must involve the end users and the specialist suppliers and must include the use of value planning to define and record the end user's detailed functional requirements and values, and value engineering to validate the design and construction options against those end user requirements and values.

There are excellent and detailed guides on value management available in the UK that can support the structured approach to defining the end user's requirements and values. In simple terms the process is as follows:

1   The value management facilitator identifies those organisations and those individuals that need to be involved in the value management workshop (or workshops for a larger project). This must include the identification and inclusion of those key end users and specialist suppliers that will have the greatest impact on defining and delivering the values of the project. Those individuals in the end user organisations that are able to take a long-term view of the way that functional requirements and values might change over time are particularly critical to the value management process since they will be able to ensure that a degree of flexibility is incorporated.

2   One or more formal workshops are held to tease-out the full range of detailed end user requirements and values. These must be precisely defined and listed in order of importance and in terms that relate to the business needs of the end user. They must not be expressed in terms that relate to the constructed solution, for example, in the case of a sports hall, the total range of activities that must be accommodated by the completed sports hall must be defined and recorded, not the shape or size of the hall. If some activities are to be to an international

standard and others to a local standard, this must be captured and recorded so that there is absolutely no possibility of ambiguity or confusion.

3   The detailed, comprehensive and unambiguous end user functional requirements and values must be expressed in simple, easy to understand language that every end user can understand, and must be signed-off by all end users to ensure there is no possibility of conflict or disagreement when the facility becomes available for use.

4   The developing design must be validated against the end user functional requirements in formal value engineering workshops. Again, these will be most effective if attended by those individuals that attended the initial value management workshops. This is especially true of the end users, since they will then understand why the design decisions were made, will have been involved in the selection of the preferred options and will consequently feel a sense of ownership for the built solution. It is essential that the value engineering process be seen as enhancing the whole-life value, not merely reducing the initial capital cost. It is also essential that the workshop pays as much attention to the labour element as it does to the materials or component element, since the labour element may be greater than the materials element and may offer the greatest potential for savings in both capital and whole-life terms.

5   Finally, the functional performance of the completed facility must be reviewed and compared with the initial user functional requirements to ensure that the end users have received every aspect of their detailed and precise business needs.

A good gauge to the success of the value management/value engineering process is to speak to the specialist supplier's operatives on site. If the carpenter or the electrician understands and can describe in detail the functional requirements and values of the end users, and if they associate their success with the ability of the functional performance of the completed facility to delight the end users, then the value management/value engineering process has been a success.

9 Selecting the independent experts

In earlier chapters, the importance of demanding evidence from supply-side firms to back up claims of improvement in performance that prove their working practices accorded with the best practice of the three standards (*Charter Handbook*, *Modernising Construction* and *Better Public Buildings*) was strongly emphasised. This applies equally well to the selection and appointment of the independent experts who are essential to supporting the change process within any demand-side or supply-side firm or organisation as advisers, trainers, coaches and mentors. There is little point spending money hiring experts whose knowledge and experience of the many different aspects of the best practice of the three standards is not much more than your own. Never assume that what is being claimed is what is actually known or what has actually been experienced, always insist on hard evidence to back up what is being claimed.

Case history: Appointment of lean construction experts

The old adage, 'caveat emptor – let the buyer beware (he alone is responsible if he is disappointed)', is very apt to the selection of expert consultants. As an illustration of the very real risk the construction industry buyer of expertise runs, during the early part of 2002 a major UK organisation was seeking experts who could advise on 'lean construction'. Several consultants who claimed expertise in this field were invited to a meeting and each was asked to define and explain the term 'lean construction'. Unfortunately, each gave a totally different definition and none of the definitions nor the associated explanations accorded with the organisation's understanding of 'lean' from their knowledge of the manufacturing sector. It was therefore assumed that those attending the meeting were not the experts they claimed to be, but were exploiting a superficial knowledge of 'lean construction' to open up a new market and were hoping their superficial knowledge was greater than that of their potential client. Needless to say, such a low level of knowledge in a so-called expert would be a major barrier to improvement in the change process.

Where the independent expert is required to assist and support a change process that relates to the best practice of the three standards, it would be logical to use the two key differentiators and six primary goals of best practice from the Construction Best Practice Programme booklet, *A Guide to Best Practice in Construction Procurement*, as the evaluative criteria of the selection process. The six goals would be especially effective in teasing out the relevant evidence and would ensure the independent expert really did have the depth and span of knowledge and experience that would enable him to adequately and constructively support the drive for improvement.

The evidence sought against each of the six primary goals of best practice in construction procurement ought to include the following:

Functionality

- What evidence can they provide of active involvement in both value management and value engineering workshops?
- Have they read the three standards (*Charter Handbook*, *Modernising Construction* and *Better Public Buildings*), and can they briefly summarise their key aspects?

- Can they give some of the examples cited in *Better Public Buildings* to illustrate the critical importance of delighting the end users?
- Can they explain how maximum functionality would benefit the competitiveness of an end user?
- Could they briefly set out an action plan for ensuring maximum functionality was delivered?

Cost of ownership

- Have they read the Confederation of Construction Clients' publication, *Whole Life Costing – A Client's Guide*, and can they briefly summarise the main points of the guide?
- Have they read and used the *Component Life Manuals* produced by the Building Performance Group (which is linked to the Housing Association Property Mutual insurance company) and published by Spon Press and Blackwell Publishing, and can they briefly summarise the purpose and value of the manuals?
- Can they briefly explain why predicting the cost of ownership would benefit the end user's business planning process?
- Can they explain the difference between predicting the cost of ownership and estimating the cost of ownership?
- Can they explain why the *Charter Handbook* is concerned about defects during the usage of a building and how these affect the end user's competitiveness?

Inefficiency and waste

- Have they any direct experience of measuring the effective utilisation of labour and materials?
- Can they briefly describe what the 1994 Latham Report had to say about the level of inefficiency and waste in the construction industry?
- Can they briefly describe what the National Audit Office Report, *Modernising Construction*, had to say about the degree to which the industry measures the effective utilisation of labour and materials?

- Can they briefly explain why it might be important to measure the effective utilisation of labour and materials?
- What do they believe the average level of the effective utilisation of labour and materials to be?

Specialist suppliers

- Can they briefly describe the benefits the involvement of specialist suppliers from the outset could bring to the design process?
- Can they offer any evidence of direct experience in the early involvement of specialist suppliers in design?
- Have they read the Reading Construction Forum publication, *Unlocking Specialist Potential*, or the Building Services Research and Information Association Technical Note 14/97, *Improving M and E Site Productivity*?
- Can they suggest how the early involvement of specialist suppliers could deliver 'right first time' on site?

Single point of contact

- Can they give any reasons why a single point of contact for the procurement of a building might aid the integration of design and construction?
- Can they briefly describe the likely barriers to procurment through a single point of contact?
- Can they explain how a single point of contact would support supply chain management?
- Have they any direct experience of single point of contact procurement?
- Can they briefly explain why single point of contact procurement is favoured by all three standards (*Charter Handbook*, *Modernising Construction* and *Better Public Buildings*)?
- Can they cite any form of traditional procurement that would ensure the involvement of specialist suppliers from the outset of design development?

Measurement

- Can they explain why all three standards see measurement as key to improvement?
- Can they explain why measurement is fundamental to an effective improvement programme?
- Can they give an example of an existing measurement system that has been tested and approved by major repeat clients?

None of the above questions has a precise right or wrong answer, but a comparison of the responses should separate out the true experts who will have based their responses on a thorough understanding of the best practice of the three standards and on a real and direct experience of every aspect of supply chain integration and management.

When selecting experts to support the improvement in performance that is demanded by the best practice of the three standards, it should always be borne in mind that the primary key to radical improvement in the construction industry is the elimination of the high level of inefficiency and waste in the utilisation of labour and materials. This was first highlighted in the 1994 Latham Report, *Constructing the Team*, which estimated that 30 per cent of the initial capital cost of construction was consumed by inefficiency and waste in the utilisation of labour and materials. The Egan Report, *Rethinking Construction*, (as do the *Charter Handbook*, *Modernising Construction* and *Better Public Buildings*) believed the elimination of inefficiency and waste to be of critical importance and also saw supply chain management techniques and the introduction of supply chain integration as key to the early involvement of specialist suppliers. The Singapore *Construction 21* Report saw buildability as key to the elimination of inefficiency and waste in the utilisation of labour and materials, and recognised that this could only be achieved by harnessing the skill, knowledge and experience of the specialist suppliers from the outset of the design process.

It should also be borne in mind that if effective supply chain management is to be delivered, it is self-evidently true that the effective management must apply to the whole of the design and construction supply chain. Effective management cannot be achieved with split responsibility because those supposedly being managed can play one manager off against the other. Split responsibility also has the very serious weakness that no single individual is ultimately and solely responsible for the performance and outputs of the total team. It follows from this that effective supply chain management requires a single point of contact for the client and this must be formalised in the client's contract for the procurement of the building or facility.

To be of any real value, the independent expert must fully accept the failings of all forms of current procurement that are described in detail in the three standards, particularly the National Audit Office Report, *Modernising Construction*. The independent expert must also accept the validity of the insistence in all three standards that the solution to the industry's failings is to be found in the importation of supply chain management techniques from other sectors particularly from the manufacturing sector. This definition of the failings of the supply-side of the construction industry and of the solution that must be adopted go back to the 1994 Latham Report and 1998 Egan Report; the three standards merely reinforced and made more explicit what was wrong and what had to be done.

10 The benefits of best practice and the risk of ignoring change

This book has set out the very real and undeniable commercial benefits for both end users and supply-side firms that can be gained from adopting the best practice approach described in the three standards (*Charter Handbook*, *Modernising Construction* and *Better Public Buildings*).

The major reduction in capital and running costs considerably improved functionality and durability, and greater cost and time certainty will be of major commercial benefit to clients and end users. These benefits will also be of great advantage to PFI/PPP (Private Finance Initiatives/Private and Public Partnerships) contractors and Special Purpose Vehicles, since they would increase their competitiveness. They would also enhance their profits and assure them over the lifetime of the PFI/PPP contract by reducing the long-term risks that come from premature failure of components or materials.

Harnessing the skills, knowledge and experience of the specialist suppliers into design from the outset will bring massive commercial benefits to every firm in the design and construction supply chain. It will end reworking, maximise standardisation of components, materials and processes, maximise the effective utilisation of labour on site, minimise risk (including cost of ownership risk), and it will boost the morale of every individual in every firm in the supply chain. These benefits will enable the high unnecessary cost generated by inefficiency and waste in the utilisation of labour and materials to be converted into higher salaries and wages, more research and development, more training and lower prices to customers.

The commercial benefits that come from total integration and management of the supply chain and the concomitant elimination of unnecessary cost, has been demonstrated time and again in other sectors and the construction industry will be no exception to that rule. As a consequence, those supply-side firms embracing best practice will increase their market share and share value. The following case histories have already been used to illustrate the benefit of supply chain management in earlier chapters, but the CRINE example makes them worth repeating to reinforce the message of this chapter.

Case history no. 1: CRINE

The above can be clearly demonstrated by a simple case study of a painting contractor from the 1996 North Sea Oil Industry CRINE (Cost Reduction in the New Era) initiative. The CRINE initiative was necessitated because oil prices slumped from $23 a barrel down to $13 a barrel and the capital and operational costs of oilrigs were a major factor in the cost of oil production in the North Sea. The solution agreed by all the oil firms operating there was for individual oil companies to work closely with their specialist suppliers within an integrated supply chain, building on the trust generated by long-term, strategic supply chain partnerships. This enabled them to use supply chain management techniques to harness the considerable knowledge and experience of the specialist suppliers to drive out unnecessary costs.

The price for painting an oilrig prior to the introduction of supply chain management was £4 million. His profits were varied, frequently entailed a loss and quite often came from claims. His buyers bought the cheapest paint they could find and he paid his operatives as little as he could get away with. After the introduction of supply chain management, his price for painting an oilrig was £2.8 million, but he made a very comfortable and very assured profit. His buyers now worked very closely with paint manufacturers to buy high quality paint with excellent durability and applicability characteristics. He employed fewer operatives, but he paid them a far higher wage in return for an end to reworking.

Case history no. 2: Building Down Barriers

The Building Down Barriers test-bed pilot projects demonstrated precisely the same commercial benefits. In the case of a steel fabrication firm, the input of the skill, knowledge and experience of the erectors and fabricators into design development from the outset ensured that the steel frame was erected 'right first time' on site with no reworking and no wastage of components or materials. This produced a much lower price for the client, but it also produced much higher profits for the steel fabrication firm. It also boosted the morale of the erectors and fabricators because, for the first time, they were being properly valued and given the opportunity to make a very positive, direct contribution to the design of the building. Virtually all the firms in the supply chains of the two Building Down Barriers test-bed pilot projects had the same experience and were convinced of the massive commercial benefits that would come from adopting true supply chain integration and the concomitant six primary goals listed in Chapter 3.

The intense external pressure on clients to adopt the three best practice standards will inevitably create a two-tier industry. Those supply-side firms that adopt the approach described in the three standards will be able to provide the well-documented evidence that clients will increasingly demand. As a consequence, they will increase their market share and enhance their share value at the expense of those firms that have refused to embrace the best practice of the three standards (*Charter Handbook*, *Modernising Construction* and *Better Public Buildings*). In addition, adoption of the best practice of the three standards will deliver higher and more assured profits by converting the unnecessary costs due to the inefficient utilisation of labour and materials into profits.

Conversely, those supply-side firms that ignore the best practice approach to construction procurement described in the three standards will find it increasingly difficult to sell their services. This is because they will be unable to provide the well-documented evidence of improvement in performance required by the growing number of public and private sector clients adopting the approach to best practice described in the three standards.

The use of *Modernising Construction* and *Better Public Buildings* as the evaluative criteria for the regular external audits of public sector bodies by the National Audit Office (who audit central government bodies) and the Audit Commission (who audit health and local government bodies) will force the rapid adoption of the three best practice standards across the entire public sector. This reality is amply demonstrated by the requirement imposed on Housing Associations in the spring of 2002 to accord with the procurement standard set in the *Charter Handbook* as a condition of the loan to fund the houses.

As the public sector is responsible for 40 per cent of the total annual construction expenditure in the UK, any radical change in public sector procurement practice will create a very real threat to those supply-side firms that fail to adopt the best practice approach described in the three documents. Their market share will fall, as will their share value. They will lose their best staff to the supply-side firms that show the foresight to set aside their inefficient, fragmented and adversarial practices and processes and adopt the efficient, integrated approach to design and construction described in the three standards. This migration of the best staff is inevitable, since the obvious source of experienced staff for those supply-side firms that are thriving and expanding because they have adopted the best practice of the three standards will be from those supply-side firms that are struggling to survive because they refused to react to a changing and more knowledgeable market.

Survival in the future will depend entirely on the adoption of best practice as defined in the three published standards. Survival will also depend on the speed at which real change can be embraced and used to provide the well-documented evidence which the clients (and their external auditors) will require to prove to the firms with which they are proposing to contract to procure that a constructed product can deliver better value for money.

In the case of the increasing number of end user clients that will adopt the best practice of the *Charter Handbook*, *Modernising Construction* or *Better Public Buildings*, this evidence is likely to be assessed in terms of the six primary goals of construction best practice, namely:

1   That the finished building will deliver maximum functionality, including delighted end users.
2   That end users will benefit from the lowest cost of ownership.
3   That inefficiency and waste in the utilisation of labour and materials will be eliminated.
4   That specialist suppliers will be involved in design from the outset to achieve integration and buildability.
5   That the design and construction of the building will be achieved through a single point of contact for the most effective co-ordination and clarity of responsibility.
6   That the current performance and the improvement achievements will be established by measurement.

11 Further reading

For a short, plain English explanation of **best practice procurement**:

A Guide to Best Practice in Construction Procurement
Available from the Construction Best Practice Programme, PO Box 147, Watford WD25 9UZ. Telephone: 0845 605 5556, Email: helpdesk@cbpp.org.uk, Website: www.cbpp.org.uk

A simple guide aimed at working-level staff in all sectors of the industry, from end users to manufacturers. It explains the historical background to the 'Rethinking Construction' movement and briefly describes the key aspects of the three best practice standards (*Better Public Buildings*, *Charter Handbook* and *Modernising Construction*). It sets out the next step actions that each sector must take in order to achieve the best practice of the three standards and warns of the consequences of doing nothing.

For detailed descriptions of **best practice construction procurement**:

Better Public Buildings
Available from the Department of Culture, Media and Sport, 2–4 Cockspur Street, London SW1Y 5DH. Telephone: 020 7211 6200, Website: www.culture.gov.uk/pdf/architecture.pdf

A short, lucid guide (six pages of text) that focuses strongly on the business benefits of well-designed buildings that enhance the quality of life, and therefore the efficiency, of the end users. It also explains the business benefits of using whole-life costs as the basis of design and construction decisions, and it makes clear that best practice necessitates the appointment of integrated design and construction teams.

The Clients' Charter Handbook
Available from the Confederation of Construction Clients, 1st Floor, Maple House, 149 Tottenham Court Road, London W1T 7NF. Telephone: 020 7554 5340, Fax: 020 7554 5345, Email: cccreception@ccc-uk.co.uk, Website: www.clientsuccess.org

A short, lucid guide (twelve pages of text) that explains the approach to construction procurement that every chartered client must adopt. It focuses strongly on the importance of the client's leadership role within an integrated design and construction supply chain, which targets major reductions in whole-life costs, substantial improvements in functional efficiency and the elimination of defects over the whole life of the building. It also emphasises the benefits to repeat clients of long-term, partnering relationships with all key suppliers.

Modernising Construction
ISBN 0–10–276901–X
A report by the Controller and Auditor General of the National Audit Office (NAO) and available from any Stationery Office bookshop or by contacting NAO. Telephone: 020 7798 7400, Email: enquiries@nao.gsi.gov.uk, Website: www.nao.gov.uk/publications/

A very comprehensive report which sets out in detail the many barriers to improving construction industry performance and describes the various industry initiatives since 1994. It concludes that better value means better whole-life performance and this can only come from total integration of the design and construction supply chain through a single point of contact. This ensures the involvement of the specialist suppliers in design from the outset, which is key to the elimination of inefficiency and waste, the achievement of optimum whole-life costs and the delivery of maximum functionality.

*Local Government Task Force (LGTF) 'Rethinking
Construction Toolkit'*
Available from the Customer Sales Department, Thomas Telford
Ltd, Unit 1/K Paddock Wood Distribution Centre, Paddock Wood,
Kent TN12 6UU. Telephone: 020 7665 2464, Fax: 020 7665 2245,
Email: orders@thomastelford.com

This provides local authorities with a valuable support to the
abandonment of outdated procurement practices that cause waste,
in terms of the inefficient use of labour and materials, poor
whole-life performance and poor functionality. It provides simple,
practical 'How To' guidance that will enable local authority
staff to introduce 'smart' procurement as recommended by the
Egan Report.

For detailed guidance on **supply chain integration and management**:

*The Building Down Barriers Handbook of Supply Chain
Management – 'The Essentials'*
ISBN 0–86017–546–4
Available from the Construction Industry Research and Informa-
tion Association (CIRIA), 6 Storey's Gate, Westminster, London
SW1P 3AU.

An overview of the Building Down Barriers approach to supply
chain integration and an introduction to the toolset as a whole.
It describes the seven underlying principles of total supply chain
integration and the lessons learned from their application on the
two test-bed pilot projects. It also describes the benefits and
the challenges of supply chain integration for the various sectors
of the industry.

For guidance on **predicting whole-life costs**:

Whole Life Costing – A Client's Guide
Available from the Confederation of Construction Clients, 1st
Floor, Maple House, 149 Tottenham Court Road, London W1T

7NF. Telephone: 020 7554 5340, Fax: 020 7554 5345, Email: cccreception @ccc-uk.co.uk Website: www.clientsuccess.org

A short, lucid guide (nine pages of text) that explains to clients the benefits, in business planning terms, of making construction investment decisions on the predicted cost of ownership. It explains the level of accuracy that can be expected at the various stages of design and construction and makes clear that optimum whole-life costs can only be achieved with the early involvement of specialist suppliers in design.

Technical Audit of Building and Component Methodology
Available from the Building Performance Group, Grosvenor House, 141–143 Drury Lane, London WC2B 5TS. Telephone: 020 7240 8070.

Describes a technical audit process for assessing the whole-life performance of buildings and can be used as a first, second or third party audit system.

HAPM Component Life Manual
Available from Spon Press, Cheriton House, North Way, Andover, Hampshire, SP1O 5BE. Telephone: 01264 342933.

The *Manual* schedules over 500 components and gives the insured life, maintenance requirements and adjustment factors. The insured lives are cautious, were developed for housing, and are limited to 35 years, so should not be used without adjustment. The *Manual* is updated twice a year, contains references to current British and European Standards and includes feedback from research and claims on HAPM latent defect insurance.

Building Services Component Life Manual – Building Lifeplans
Available from Blackwell Science, Osney Mead, Oxford, OX2 0EL. Telephone: 01865 206206, Fax: 01865 721205.

This *Manual* provides much needed guidance on the longevity and maintenance requirements of mechanical and electrical plant. It sets out typical lifespans of building service components – boilers, pipes, ventilating systems, hydraulic lifts, etc. These are ranked according to recognised benchmarks of specification, together with adjustment factors for differing environments, use patterns and operating regimes. Summaries of typical inspection and maintenance requirements are provided, along with specification guidance and references to further sources of information.

For detailed guidance on involving **the specialist suppliers in design**:

Unlocking Specialist Potential
ISBN 1–902266–00–5
Available from Reading Construction Forum, PO Box 219, Whiteknights, Reading, Berkshire, RG6 6AW.

A detailed guide that explains how the skill and experience of specialist suppliers can be harnessed in design development. It proposes strategies for better teamwork and collaboration; for a process-orientated approach to design and construction; and for a central focus on customer requirements. It makes clear that it is only by enabling the specialist suppliers to play a key role within the design process that real improvements in value can be achieved. The guide was used to develop the technology cluster concept in the Building Down Barriers toolset.

For detailed international evidence of **labour inefficiency levels**:

BSRIA Technical Note 14/97 'Improving M & E Site Productivity'
Available from the Building Services Research and Information Association, Old Bracknell Lane West, Bracknell, Berkshire, RG12 7AH. Telephone: 01344 426511, Fax: 01344 487575, Email: bookshop@bsria.co.uk, Website: www.bsria.co.uk

Comprehensive evidence from projects in UK, USA, Germany, France and Sweden on the true levels of the efficient use of labour and the causes of the inefficiencies described in the report, including naming the sector of the industry that was responsible for the individual problem. It also gives advice on how to improve efficiency levels by better integration and co-ordination.

For measurement of **effective labour utilisation and training in best practice procurement:**

CALIBRE The productivity toolkit
For further information contact the Centre for Performance Improvements in Construction (CPIC), BRE, Garston, Watford, Hertfordshire WD2 7JR.

CALIBRE provides a consistent and reliable way of identifying how much time is being spent on activities that directly add value to the construction and how much time is being spent on non-added value activities.

For **training and coaching in best practice procurement:**

ICOM/CITB Diploma in Construction Process Management
For further information contact ICOM, Long Grove House, Seer Green, Buckinghamshire HP9 2UL. Telephone: 01494 675921, Fax: 01494 675126, Email: Kinder.ICOM@btinternet.com

ICOM is linked with the Construction Industry Training Board (CITB) and the University of Cambridge Local Examinations Syndicate and uses the Construction Best Practice Programme booklet, *A Guide to Best Practice in Construction Procurement*, to define best practice in its training. ICOM is also working with the CITB to offer awareness workshops and coaching for clients on best practice construction.

For **advice on possible integrated teams:**

Design Build Foundation
Available from PO Box 2874, London Road, Reading RG1 5UQ. Telephone: 0118 931 8190, Fax: 0118 975 0404, Email: enquiries@dbf.uk.com, Website: www.dbf-web.co.uk

The Design Build Foundation was incorporated in 1997 with the aim of bringing together representatives from the whole construction industry to improve the integration of design and construction and to deliver customer and team satisfaction through a single source of responsibility. It is a self-funded, multi-discipline organisation comprising leading construction industry clients, designers, consultants, contractors, specialists, manufacturers and advisers. It is not a trade organisation for design and build companies and does not support 'old design and build' practices.

Index